How To Protect Your Pets, Livestock and Home Garden In A Nuclear Event

By Charles S. Brocato (Dr. "B")

Text and Images © 2018 Charles S. Brocato except where otherwise noted.

All rights reserved

http://www.chemicalbiological.net

Cover Photo: Alex Profa/shutterstock.com

All Other Photos: Charles S. Brocato

Dedicated to:

Albert J. Love, M.D.

&

J.M. Stark, Ph.D., Mathematics

Acknowledgements

Thank you to Dr. Al Wehner, of Calder Animal Clinic in Beaumont, Texas, for his opinions and guidance on protecting pets with KI during a nuclear incident.

Many thanks to Kathryn E. King, who aided me with every aspect of this book. Without her encouragement and help, this project might never have been completed.

Table of Contents

Acknowledgements 4

Introduction 7

1: *The Threat To Pets In A Nuclear Event* 9

2: *Why Is Fallout Dangerous?* 17

3: *Some Isotopes In Fallout* 29

4: *Your Radiological Instruments* 39

5: *Your Home Fallout Shelter* 51

6: *How To Prepare For Fallout* 57

7: *What If Fallout Arrives Unexpectedly?* 63

8: *Protecting Your Pets* 71

9: *Handling Contamination* 75

10: *Using Your Animal Products* 85

11: *Restoring Your Garden* 89

12: *Radiation Effects on Insects & Worms* 95

Radiation Safety Limits 99

Radiation Limits For Outside Activities 101

Resources 103

About the Author 107

Other Books By This Author..109

Introduction

Charles S. Brocato, known to his students as Dr. "B," began writing a series of books on nuclear radiation and what you can do to protect yourself in 2017. In Book I of his radiation series, he showed you how to choose the radiological instrument(s) you need to know the levels of radiation around you if a nuclear event should occur.

Book II discussed how you can construct and stock your own home fallout shelter, which is now a necessity because the government stopped the public fallout shelter program in the 1970s. There's nothing out there for the American people any longer.

In Book III, he showed you what nutrients you can use to help reduce any damage to your body from radiation.

This book, Book IV of Dr. "B"s Radiation Series, examines what steps you can take to protect your pets and livestock from nuclear radiation in the event of a nuclear bomb or a nuclear plant accident.

In America, pets are often beloved members of the family, and many city households nowadays have a home chicken flock. Even farmers or homesteaders who may have dozens of livestock, are concerned with their animals' wellbeing and are willing to go to great lengths to protect their animals.

Dr. "B" studied the research done during the Cold War, when authorities feared that a nuclear attack on America would affect agriculture and livestock and drew on his own vast knowledge of radiation and its effects, nutrition and biochemistry, to show you what you can do to protect your pets, your livestock, and even your home vegetable garden, in the event of a nuclear incident. This book is the result of his research.

1: *The Threat To Pets In A Nuclear Event*

Once upon a time, everyone believed nuclear war with the Soviet Union would occur any day, so Americans, both the government and the people, prepared accordingly. School children were taught to *Turn, Duck and Cover*, or to dive beneath their desks in the classroom. Parents were instructed on how to stock their home fallout shelters or what to bring to a public fallout shelter. There was even a Conalrad radio system to alert and inform the people if an attack should occur.

Then came the days of "Mutually Assured Destruction," when the government quit all programs designed to protect the American people and allowed everyone to believe that in a nuclear war, everyone was dead anyway, so why should you do anything to try and survive all by yourself?

The truth, of course, is that most people not at or extremely near Ground Zero will survive a nuclear attack. Many people today, in spite of historical and common-sense evidence, still cling to this belief.

Many farmers and homesteaders who live far away from city targets, with livestock and poultry to protect, will also survive.

So will many city folks, even those within a "danger zone," who have cats and dogs they love and consider members of the family. Many city dwellers these days also keep a small flock of laying hens, each one of which is named and cherished.

Some city or suburb dwellers may even have a milk cow, or a milk goat or two. Miniature pigs are considered by many the finest, most intelligent pet available. When it comes to pets, many city folks like to keep certain farm animals. In fact, some city property has been turned into miniature farms, complete with gardens and livestock

This isn't even considering the so-called exotic pets available, like ferrets, canaries, parakeets, parrots, lizards and snakes.

We have found that all animal owners, no matter what their preference, are concerned for the animals they raise and wish to take any steps they can to ensure their animals' safety in a radiation storm.

Do not look for the government to offer you shelter, or even to give you any advice. Government these days is strictly reactive when it comes to citizen safety, not proactive. If there isn't a regulation it can enact, government is helpless.

Frankly, we will be surprised if, in the event of a nuclear attack or accident, the government even tells us the truth about the radiation levels in our area—assuming it knows. Judging from past events, such as the Three Mile Island partial meltdown in 1979, if one government agency even knows what the readings are, there will be several other government agencies that will deny those readings and give other, radically different readings.

This is why you need your own radiological meter, so you can take readings of your own shelter and the area around it. Even if the government tells the truth about readings in your area, readings can vary widely from place to place within an area. See Book I of our Radiation Series: *How To Choose a Civil Defense Radiological Instrument: Geiger Counters & Dosimeters*, for advice on which meter you need and where to purchase it.

You should also make arrangements to construct your own home fallout shelter, as described in Book II of our Radiation Series: *Your Home Fallout Shelter: How to Ensure Your*

Family's Health & Survival in a Nuclear Incident. This book also shows you what foods you need to put aside in case you and your family must stay inside your shelter for a week or two.

If you have pets or livestock, you will want to make plans to include them in your shelter, or consider dedicating a certain sheltered area to your poultry or livestock. Or you may decide to shore up the radiation defenses of your current livestock shelters. This book will help you in making those plans and decisions to protect your animals.

It has been said that once you name an animal, it is no longer just a denizen of the barnyard. The animal now occupies a place in your heart, and you will naturally want to do everything you can to make sure the creatures you are responsible for live through any nuclear event that may happen in the future.

What Nuclear Events May Threaten Your Pets?

Any event that threatens people's lives also can harm pets and livestock. Animals vary in their susceptibility to radiation, and even within a species, individual differences exist just as they do among people. Swine are similar to humans in their reactions to radiation doses, while poultry can withstand much higher doses. And within each herd of swine or flock of chickens, certain hogs or hens can take more radiation than others before showing symptoms.

With this in mind, what nuclear events are of sufficient power to threaten you and your animals, especially if you live in a suburban area or in a small town, or even far out in the countryside?

If you live in certain cities that are considered "targets" for an enemy nuclear strike, your household may experience the negative effects of a nuclear bomb blast. If you live in even a small town near a nuclear power plant, you could be affected if the plant suffers an accident. Also, your home may be located in an area that receives some debris from a "dirty bomb" blast.

If you do not live in a target city, two events can still affect you, a nuclear bomb blast and a nuclear plant accident if the winds happen to blow the nuclear debris from the blast or the nuclear plant emissions in your direction.

Remember, if a nuclear bomb is exploded over or dropped directly on your city, unless you are at or very near *Ground Zero*, the area directly beneath the bomb, you will likely survive, although you may have laceration injuries from flying debris and glass, or burns from the thermal radiation.

If you remember the old, often-mocked advice of "Turn, Duck and Cover," from the teachings of the 1950s, you may avoid the lacerations and thermal burns that injure most people who are caught near enough to Ground Zero. Contrary to popular belief, people near a nuclear bomb blast are not injured by radiation so much as by flying glass and debris from the force of the blast itself.

How A Nuclear Bomb Injures People and Animals

When a nuclear bomb explodes, a sequence of events occurs that, if you become quickly aware of what is happening, even though you may be caught outside, you can act quickly to save yourself a lot of pain and injury.

First, the atomic chain reaction sends out *initial radiation* (also called *prompt radiation*) in the form of gamma rays and neutrons that mostly affect living things that are near Ground Zero.

Second comes *thermal radiation,* in the form of heat so powerful, it vaporizes anyone near Ground Zero and sets flammable materials like houses on fire for miles outward.

Third comes the blast wave itself. This is a powerful pressure caused by the compression of the air in the vicinity of the bomb, which can be thought of as the "knock-down" wave or shock wave.

The last and most lingering effect of a nuclear blast is what scientists call *residual radiation.* This comes about when a bomb explodes on or close to the ground and vast amounts of dirt, dust and bomb debris are sucked high into the atmosphere as the initial fireball of the explosion creates a vacuum when it fades out. These grains of dirt and debris become coated with radioactive particles from the bomb itself as they roil about in the atmosphere.

Fallout

What goes up must come down ... eventually, and nuclear bomb fallout follows the pattern you would expect if you tossed a handful of sand into the air. The heavy larger particles would return to earth quickly and land on the ground fairly close to Ground Zero, where the bomb initially exploded. This is called *early fallout*, which returns to earth within 24 hours.

The medium particles might catch a wind current and travel several miles on the wind before returning to earth, and the fine, light particles might travel high into the stratosphere, not returning to earth for months, and when they do return, they may land on the ground on the other side of the globe. These particles that do not descend to the earth right away, but come down after the first 24 hours, are called *delayed fallout.* Those very fine particles that remain aloft for months to years and can come down almost anywhere are known as *worldwide fallout*.

Worldwide fallout is the fallout that is responsible for such anomalies as the "radioactive reindeer" of Norway and the radioactive wild boars in Sweden. In this case, the fine fallout from the Chernobyl nuclear power plant explosion deposited on the lichens and mushrooms the reindeer feed on in Norway, and cesium-137 is concentrated in mushrooms eaten by wild boars in Sweden. Thus, 30 years after the Chernobyl explosion, which was not even an atomic bomb explosion, fallout from the event still exists in the food supply of animals.

Worldwide fallout is also the reason fallout from the nuclear bomb testing programs of the 1940s through 1992 is found in the snows of the Arctic region and has been detected in other parts of the world far removed from the actual testing sites.

Oddly enough, the Chernobyl accident was not technically a nuclear explosion—it was a steam explosion caused by a lack of water to cool the superheated core of the nuclear reactor. The powerful heat turned what water there was into steam, which in itself is capable of generating powerful explosions.

But the steam blast itself created in the damaged reactor a "chimney effect" that caused radioactive particles from the core to be swept high into the atmosphere, so high that the prevailing winds carried what amounted to fallout to various parts of the European continent and Scandinavia.

A few days after the accident, a worker arriving for work at an atomic plant in Sweden set off the radiation alarms, leading to a frantic search for the source of the radioactive contamination. When the contamination turned out to be on the soles of the worker's shoes, authorities realized the source of contamination was the parking lot. It didn't take long before they determined the radioactive plume that dumped some of its particles on the parking lot had originated in the Soviet Union, and prevailing east winds had carried it to Sweden.

With this in mind, we can clearly see that the chief danger to most people in a major nuclear incident, especially those who live far enough away from any "target" zone to avoid the blast effects, will be wind-carried fallout.

If you do not live close enough to the site of an actual nuclear bomb explosion, or to a nuclear power plant that suffers an accident of some sort, to be directly affected by the blast effects of the bomb or an explosion at the power plant, then chances are, what is most likely to affect you and your pets or livestock is nuclear fallout.

2: *Why Is Fallout Dangerous?*

Fallout is dangerous because it is *radioactive*, meaning it contains many atoms that are unstable, which came from the bomb itself. These unstable atoms contain too many protons and/or neutrons in their nucleus and thus have too much energy, so they attempt to achieve stability by throwing off pieces of the nucleus until they achieve a less-energetic arrangement that is stable. These pieces an atom throws off from its nucleus are what we call *nuclear radiation*.

Nuclear radiation is also called *ionizing radiation* because it has the ability to knock pieces off other atoms or molecules. Some types of nuclear radiation are particles and can literally knock off pieces of other atoms like a cannon ball can knock chunks of rock off a rock cliff. Some nuclear radiation is not in the form of particles, but is a high-energy, highly penetrating ray that can also knock off components of any atoms or molecules it strikes.

Nuclear Radiation

Nuclear radiation consists of five major items: *Alpha particles, beta particles, gamma rays, x-rays* and *neutrons*. Atoms of particular radioactive elements tend to *decay*, or lose particles, in certain ways.

First, it may throw off an alpha particle. Later, a beta particle, and somewhere in between, a gamma ray may be emitted. By the time a radioactive, or unstable atom, has achieved stability, it may no longer be the element it once was. If it has lost alpha or beta particles, it has changed the number of protons it once had in its nucleus. The number of protons in its nucleus is what gives an element its name and its chemical properties.

Radioactive decay is a complex subject, and scientists have identified the sequence in which most well-known radioactive elements decay. Probably the best known radioactive element is uranium, which has a number of unstable *isotopes* (atoms that are the same element but have different atomic weights). Unstable uranium isotopes decay in certain steps in their search for stability, and their ultimate destination is to become a stable lead atom. To get there, they may have to shed quite a few particles and rays of energy, and they tend to do this in an orderly way.

The particles these unstable elements eject from their nucleuses are what we call *nuclear radiation*.

Alpha particles consist of two protons and two neutrons. They carry a positive electrical charge because of the two protons and are not particularly energetic. An alpha particle only travels a few inches in air.

It is an actual particle, one with some size and weight to it in the atomic scheme of things. For this reason, alpha particles are considered the most damaging of all nuclear radiation—if they get inside the human body via the nose, the mouth, or an open wound.

Because of their size and heft, alpha particles are not particularly *penetrating*. The epidermal layer of the skin stops them. A sheet of paper stops them. You can sit beside an alpha-particle emitter all day long and you will experience no damage whatsoever. But if you should swallow so much as a tiny amount of a vigorous alpha-emitting element like polonium, you will receive tremendous internal damage. The gastrointestinal tract has no epidermal layer to protect it, and neither do the lungs if you should breathe in a heavy dose of alpha emitters.

Because neither of these things is particularly likely in a fallout situation, alpha radiation is usually not considered a threat when you are dealing with fallout. You may accidentally inhale or swallow some alpha-particle emitters, but usually not enough to cause serious damage.

The major threat of alpha particles exists in cigarette smoke or in radon gas, and even then, the damage usually takes a number of years to show up in the form of lung cancer. Polonium-210 is present in tobacco smoke because of the high-phosphate fertilizers used on tobacco fields, which contain a lot of Radium-226. Radium-226 decays into polonium-210, which is considered the probable source of many lung cancers in smokers.

The alpha-emitter, polonium-210, was used to assassinate the former KGB agent, Alexander Litvenenko, in London in 2006, and may possibly have killed Yassar Arafat, also. Once inside the body, polonium-210 and other alpha emitting elements bombard the internal organs from within with heavy alpha particles, especially the liver, kidneys and the bone marrow. The person soon dies from multiple organ failure if the dose was heavy enough.

Alpha emitters like polonium-210 and americium-241 do not set off radiation detectors. Your radiation meters will not detect them. They require a special instrument called a *scintillator*, which gives off light rays when struck by an alpha particle. By counting the number of light flashes, you can count the number of alpha particles given off within a certain time frame.

In a fallout situation, we are usually not concerned with alpha particles unless we have to be outside while fallout is coming down. In this case, your best protection is outer clothing like a hat and a raincoat and gloves to keep fallout off your skin, and a face mask like a respirator to prevent fallout particles from entering the mouth or lungs.

Beta Particles

Beta particles are high-speed electrons given off by an overly energetic nucleus. They are more energetic than alpha particles and can penetrate the outer layer of the skin. Many radiation meters are capable of measuring beta-particle radiation.

Electrons do not inhabit the nucleus of an atom—they are found entirely in the outer shell of an atom. However, beta particles are produced when a neutron ejects an electron from inside itself and thus the neutron turns into a proton.

Beta particles are considered somewhat penetrating in that they can burn the skin if they are allowed to remain on the skin for any length of time. This is a burn you do not feel until after the burn has occurred.

Beta radiation can also do damage to your lungs if you breathe in enough fallout. Some of the brave firemen who battled the fires at the Chernobyl nuclear power plant disaster suffered huge amounts of damage to their lungs from beta burns.

Where alpha particles are heavy and capable of traveling only a few inches through the air, beta particles, with their greater energies, can travel up to six feet or more. However, a sheet of tin foil can stop them.

In a fallout situation, your chief concern will be to keep fallout from landing on your skin or head, and to avoid breathing in any fallout. This is why you want to have a raincoat, a big hat, gloves and a particle mask or respirator available if you should have to be outside in fallout.

Gamma Rays

Gamma radiation is pure energy packets ejected by a nucleus as it seeks to rearrange its energies. Because gamma radiation is not a particle and has no mass or weight, it is highly penetrating and can travel for long distances through the air. Stopping these rays requires several inches of lead or other very dense materials.

In a fallout situation, gamma radiation is usually what we are most concerned with, and all radiological instruments will measure them because this is largely what they were designed to do—measure gamma radiation in fallout, especially the old Civil Defense radiological instruments.

Gamma radiation can penetrate the entire human (or animal) body, which is why it is considered so dangerous. If the body receives enough gamma radiation, damage to the body occurs and radiation sickness results.

Neutrons

Neutrons are constituents of the atomic nucleus that add weight to the nucleus but are electrically neutral. Because they are electrically neutral and do not veer toward an opposite electrical charge, they are considered very penetrating.

Neutrons are also useful in that they can cause atoms they strike to split, which is known as *fission*. Rather than cause molecules to ionize, they crash into them and knock off pieces. Inside a nuclear reactor, neutrons drive the fission process.

After an atomic bomb explosion, neutrons often strike other atoms and either break them apart, or they are *captured* by a nucleus of a nearby atom. When this happens, the atom that captured the neutron may itself now may have an unstable nucleus that will undergo nuclear decay and release particles or energy packets of its own. This happens often in soil and rocks that lie around the area of a nuclear explosion.

Neutrons are the only type of radiation that can make other materials radioactive.

X-Rays

X-rays are very similar to gamma rays, although generally less energetic. The dividing line between the two is where they originate. Gamma rays are emitted from the nucleus of an atom, whereas X-rays originate from within the electron clouds of atoms when fast electrons strike some-thing and come to a halt, and energy is released in the form of X-rays. X-rays are also given off during atomic bomb explosions.

Which Disaster Produces What?

Both a nuclear plant disaster and an atomic bomb explosion produce radioactive isotopes that decay and produce nuclear radiation. However, each of these events produces a slightly different mix of isotopes. This is why Swedish scientists soon realized the radioactive material appearing in Sweden in 1986 had come from a nuclear plant rather than an atomic bomb test.

A nuclear plant incident releases both cesium-134 and cesium-137. Cesium-134 is formed by neutron bombardment of stable cesium-133 in the long-term environment of the nuclear reactor.

An atomic bomb produces all the same isotopes, but very little cesium-134 because the bomb explodes immediately. There is no ongoing neutron bombardment to form cesium-134 in any appreciable amounts.

An atomic bomb blast sends dirt and dust and bomb debris high into the atmosphere, from whence it begins falling back to earth, heavier particles first.

A nuclear plant accident usually releases what is called a *radioactive plume*, a mass of radioactive airborne debris and isotopes that spread from a source, namely the nuclear power plant. The plume generally behaves more or less like smoke from a smoke stack, spreading wider as it moves away from the source.

But to you and your animals, the mix of isotopes, or whether the radiation comes from a nuclear plant accident or a bomb blast does not really matter. What matters is how heavy the fallout is and the fact that heavier fallout means more radioactivity that can harm you and your pets or livestock.

For the purposes of this book, we will consider radioactive plumes and nuclear bomb fallout the same in terms of what you must do to protect yourself and your animals.

Fallout and Your Home

The big problem with fallout is that it travels where the wind goes, which means it can travel almost anyplace, and it can sometimes land where it is not expected. (This is called contamination—radioactive matter found where it is not wanted.)

Studies done back in the days when we expected an attack at any time by the U.S.S.R. show the entire United States being almost covered by fallout within 24 hours of multiple atomic explosions because of the prevailing winds. Fallout can sometimes travel several hundred miles on strong winds.

It can be hard to predict where fallout will land because winds in the upper levels of the atmosphere often travel in different directions than the winds in the lower levels. You may think the fallout is coming directly to you if you are downwind of a certain city, only to discover it went another direction entirely. Or it may come to you when forecasters said it was likely to travel in the opposite direction.

Remember the trouble with one of the Bikini Atoll nuclear tests in 1954, when the winds near a U.S. nuclear test shifted unexpectedly and carried the fallout straight to the inhabited Marshall Islands. Many Marshall Islanders received beta burns to their skin and scalps because they had not been warned and did not know what the fallout was, or that it needed to be removed as fast as possible. Wind direction can shift abruptly and with no warning, so prepare for fallout if you should hear of a nuclear explosion anywhere near you.

Add to this other possible weather problems, like rain or snow. Any meteorological events like these can cause an unexpected buildup of fallout in certain areas as precipitation will carry it down to the ground faster than it would otherwise have come down.

It can be very difficult to predict when and where fallout, or a radioactive plume, will decide to come back to earth. But when it does, we must find ways to deal with it.

How Long Does Fallout Remain Dangerous?

About the only good thing about radioactive fallout is the fact that it decays. All those radioactive isotopes continue trying to reach less energetic states by casting off alpha and beta particles and gamma rays and neutrons. This means that as time passes, the radiation given off by the fallout weakens.

Generally, if only one bomb blast sent out the fallout, most of it has decayed within a week to two weeks to what is considered "safe" radiation levels. This is why most fallout-shelter stays were estimated at no more than two weeks.

However, if there are multiple bomb blasts, especially at varying times, the fallout can keep on coming.

Radioactive isotopes each have their own method and time table for decay. Some isotopes, like carbon-15, decay within a few seconds. Others may not decay for many years. We measure radioactive decay in terms of *half-life*, the time it takes for one-half of an amount of a certain isotope to decay. Carbon-15 has a half-life of 2.449 seconds. Plutonium-239 has a half-life of 24,100 years.

For a list of radioactive isotopes and their half-lives, many of which you probably never knew existed, go here:

https://en.wikipedia.org/wiki/List_of_radioactive_isotopes_by_half-life

Some of the isotopes of interest to us in a fallout situation are the following:

Iodine-131, half-life of 8 days

Cesium-137, half-life of 30 years

Strontium-90, half-life of 28.8 years

If the fallout comes from a radioactive plume emanating from a nuclear plant, we are also interested in cesium-134, which has a half-life of 2.065 years.

These isotopes and their half-lives become important to us as we make plans to protect ourselves, our pets and livestock and our home gardens.

But the main thing to remember is that within a week or two, many of the shorter-lived isotopes have decayed, and the overall amounts of radiation being given off by the fallout has fallen considerably.

How Does Fallout Harm Animals?

Fallout harms animals in much the same way it does humans. If an animal receives a big enough dose of radiation, the animal may develop radiation sickness and even die. Different animals present varying symptoms according to their species. An animal such as a cow or sheep with serious radiation sickness may develop lethargy, bloody diarrhea and bloody discharge from the nose or mouth.

One of the biggest dangers to range animals like cows and sheep occurs when fallout lands on their backs. An animal may receive severe beta burns along its back if fallout remains there for some time. The same could happen to your pets or poultry if they happen to be outside at the time fallout comes down.

There is also a danger of your pets or animals drinking contaminated water while fallout is coming down into the water, or of your grazing livestock eating grass that newly fallen radioactive fallout has landed on. Then there is a danger of the animals receiving an internal dose of radiation, and, in the case of dairy cows or goats, of producing milk contaminated with radioactive iodine. Other products from your animals may also be affected.

How Does Fallout Harm Your Garden?

Fallout can harm gardens in several ways.

First, the beta particles can actually burn plants in the places where it may tend to collect on a plant.

Second, radioactive isotopes on a plant may be actually taken into the plant, so that whoever or whatever eats the plant receives an internal dose of radiation.

Third, when fallout particles land on your garden soil, rain and tilling may carry them deeper into the soil. There, certain radioactive isotopes may be absorbed into the plants by their root systems.

However, if you have advance notice that a radioactive plume or fallout may soon arrive in your area, you may have time to take some preventative actions.

3: *Some Isotopes In Fallout*

Fallout, whether it comes from a nuclear bomb blast or a nuclear power plant accident, is composed of a mixture of radioactive isotopes and debris. Around 200 different isotopes of almost 35 elements have been noted in the fallout created by nuclear weapons. Nuclear reactors have a slightly different mix of isotopes due to the unique environment inside reactor cores, but many are the same as those created by an atomic bomb explosion.

As we have seen, each of these isotopes has its own sequence of decay, and each has its own half-life. Most are of interest to us because no matter where they land, they produce radiation that is dangerous to human and animal life.

Others are of interest to us because they can be taken into the human body by various means, or they can taken up by plants or animals and thus into the human body when we eat those plants or animals. Some are of interest to us because they can make their way into certain animal products, such as milk, when a cow grazes on grass covered with fallout. For this reason, we need to know about them and how they get into humans, animals and foodstuffs.

Isotopes Hazardous To You & Your Animals

Isotopes we consider the most important to humans and animals are those which (1) are extremely long-lived and (2) can get into our food supply. The long-lived ones stick around in our environment and irradiate us for a lot longer than we'd like, and the ones that can get into our food supply are of even greater concern because they irradiate us from the inside.

Some of the isotopes of most concern to us, because they get into our food supply, are as follows:

Barium-140: Half-life of 12.75 days. Beta emitter.

Barium-140 is a fission product that was released in large amounts into the stratosphere (which means it can spread worldwide!) and was taken into the food chain because its chemical behavior is similar to that of calcium. Barium-140 appeared in some U.S. milk samples in June, July and August of 1957, probably because of fallout from Operation Plumb-bob in Nevada, and perhaps from U.S.S.R. nuclear testing. Although its chemical behavior is similar to that of calcium, and thus to strontium-90, and it can be deposited in bone, it is considered less dangerous than strontium-90 because of its short half-life.

Cerium-144: Half-life of 285 days. Beta emitter.

After the above-ground nuclear testing in the U.S., cerium-144 was found in food and animal bone in Japan, and clams had the highest levels. American wheat and flour were contaminated with cerium-144. After the Chernobyl disaster, many human respiratory problems were caused by an aerosolized form of cerium-144, and large particles and other debris were found in Bulgaria, Finland, Germany and Hungary. It caused beta burns to emergency workers at Fukushima. Cerium-144 is often found with iodine-131 in radioactive water.

Cesium-134: Half-life of 2.06 years. Beta and gamma emitter.

Cesium-134 is mainly seen as a product of a nuclear-reactor accident, as it requires neutron bombardment to create. The presence of cesium-134 is how scientists were sure a radioactive plume of seawater in the Pacific Ocean had come from the Fukushima disaster—the cesium-134 would have decayed long ago if the radioactive isotopes had come from nuclear bomb testing in the Pacific. Cesium isotopes behave like potassium in the human body and in the soil, where it can bind strongly to clays. It substitutes for potassium in the structure of clay minerals. This means that in potassium-poor soils, cesium isotopes are available in the top layer of soil to be absorbed into plants.

Cesium-137: Half-life of 30.17 years. Beta and gamma emitter.

Cesium-137 is an isotope that is both long-lived and very common after a nuclear event. Its long half-life means it stays with us a long time in soils, where it enters the food chain through absorption by plants. Cesium-137 behaves similarly to potassium in the bodies of humans and animals. It is readily absorbed by the gastrointestinal tract of mammals and birds, from whence it is distributed in a fairly uniform matter throughout the organs and tissues of the body. It does not tend to remain long in the body, but tends to leave like potassium. While it is inside the body, however, it may cause cellular and genetic damage. It also tends to spread readily in nature because its compounds are highly soluble in water.

Iodine-131: Half-life of 8 days. Beta emitter.

This isotope is one of the most feared fission products in the environment because of its ability to concentrate in the thyroid gland, where it then proceeds to kill thyroid tissue. It is also a major fission product of uranium and plutonium, which means both nuclear reactor accidents and atomic bomb explosions produce plenty of it. It is most dangerous during the first 60 days after a nuclear incident because of its ability to enter the food supply. It is considered highly radioactive.

Iodine-133: Half-life of 22 hours. Beta emitter.

Iodine-133 has such a short half-life you wouldn't think it would be particularly dangerous to the food supply. However, it is produced in a sufficient quantity to add to the trouble caused by iodine-133 and contributes to the total thyroid radiation dose.

Ruthenium-106: Half-life 373.59 days. Beta emitter.

Ruthenium-106 is a product of nuclear reactors and spent-fuel operations. It has medical uses in treating eye and skin tumors, and in powering radioisotope thermoelectric generators that are used in satellites. In October of 2017, a mysterious cloud of ruthenium-106 drifted over Europe that nobody would claim. It was traced to the area of the Mayak Production Association, which is known to work with spent fuel, but Russia denies any connection to it. Ruthenium-106 is toxic in high doses and is strongly retained by the bones.

Strontium-89: Half-life of 53 days. Beta emitter.

Strontium-89 behaves like calcium in the body and in the environment, which means it is taken up by plants and animals in the same way calcium would be. Although the body discriminates in favor of real calcium, enough strontium can still get into the food supply and into the bones of animals and humans to cause some real trouble.

Strontium-90: Half-life of 28.8 years. Beta emitter.

Strontium-90 is one of the big daddies of radioisotopes in that it lasts a long time and its chemical behavior is like that of calcium in soils, plants and animals. On top of that, a lot of it is formed by nuclear explosions and nuclear plant accidents, which makes it of primary importance to us. The big problem with strontium-90 is that it easily enters the food chain because plants absorb it the way they would calcium, and so do humans and animals. It tends to collect in the bones, where it can cause bone cancer. Children are more sensitive than adults because a greater amount of strontium will be deposited more uniformly in their skeletons, and it will be there much longer than in adults. After a nuclear event, it may be found in milk, cheese and animal bones. In seafood, it tends to concentrate in shells, bones and scales rather than in fish flesh.

Of these isotopes, scientists consider radioactive isotopes of cesium, iodine and strontium the most dangerous because, in addition to readily entering our food supply, they are also reasonably plentiful in the event of a nuclear incident.

Other Radioisotopes Present In Fallout

Many other radioactive isotopes are found in fallout, isotopes that do not enter the food supply, but give off dangerous radiation wherever they fall.

Actinium-227: Half-life 27.7 years. Beta emitter.

Actinium, of which there are lots of isotopes, is a decay product of plutonium-239 and uranium-235. Only actinium-227 occurs in decent amounts. This may be found in fallout from both atomic bombs and reactor accidents.

Americium-241: Half-life of 432 years. Alpha emitter, plus weak gamma rays.

Americium-241 is ubiquitous in American life today—it is the "detector" inside smoke alarms. It is a product of plutonium production inside nuclear reactors.

Americium-243: Half-life of 7,370 years. Alpha emitter.

All known isotopes of americium are unstable, since it is an artificial element formed in the operation of nuclear reactors.

Antimony-125: Half-life of 2.75 years. Beta and gamma-ray emitter.

There are many isotopes of antimony, most with very short half-lives. Antimony-125 has the longest, about 2.75 years. Many of these are produced in nuclear reactors. Antimony-124 is a strong gamma emitter.

Argon-37: Half-life of 35.04 days. Alpha emitter.

Argon-37 gas is formed when neutrons from an underground nuclear explosion bombard calcium in the surrounding rocks. Thus, its presence indicates that underground nuclear testing has taken place.

Beryllium-10: Half-life of 1,390,000 years. Beta emitter.

Beryllium isotopes are formed when cosmic rays strike nitrogen and oxygen. Beryllium-10 is also formed in nuclear explosions when fast neutrons strike the carbon in atmospheric carbon dioxide. It is an indicator of atmospheric nuclear explosions.

Bismuth-210: Half-life of 5 days. Alpha emitter.

There are many isotopes of bismuth, but bismuth-210 is a decay product of uranium-238 and is seen in the operation of nuclear reactors.

Cobalt-60: Half-life of 5.27 years. Beta and gamma-ray emitter.

Cobalt-60 tends to produce energetic gamma rays, which makes it industrially and medically useful. It is found in the radioactive wastes of nuclear power plants and in fallout from atmospheric nuclear bomb blasts.

Krypton-85: Half-life, 10.78 years. Beta emitter.

Many isotopes of krypton gas exist, but most have extremely short half-lives. Krypton gas is formed naturally when cosmic rays collide with stable krypton gas in the atmosphere. It is also produced as a fission product from atomic bomb explosions and from nuclear reactor accidents.

Lanthanum-140: Half-life of 1.68 days. Beta emitter.

Lanthanum-140 is a fission product of barium-140. During the atmospheric nuclear testing era of the cold war, it was found in snow in Central Europe after Soviet tests in 1961, and in water on the basement floor of one of the reactor buildings at Fukushima after the 2011 disaster.

Plutonium-239: Half-life of 24,200 years. Beta emitter.

Plutonium isotopes are considered among the worst isotopes released from nuclear processes, both atomic bomb blasts and nuclear reactor accidents. It is the one of the major constituents of nuclear fuel and atomic bombs. Since plutonium-239 is so long-lived, it is a major reason spent nuclear fuel remains dangerous for so long. It is found in fallout from both nuclear bomb blasts and reactor accidents.

Plutonium-240: Half-life of 6,560 years. Alpha emitter.

Plutonium-240 is formed when plutonium-239 captures a neutron. It is considered a useless form of plutonium, but it is found in fallout from both nuclear bombs and nuclear reactors.

Selenium-79: Half-life of 327,000 years. Beta emitter.

Selenium-79 is a fission product of uranium-235. It is found in fallout, in nuclear reactor spent fuel and fuel reprocessing waste. It is one of the most long-lived reactor products.

Tellurium-132: Half-life of 3 days. Beta emitter.

Tellurium isotopes are products of nuclear reactors. Tellurium-132 was detected in the area near the Fukushima reactors within about a day of the disaster, and a week later it was found in Seattle, Washington. Tellurium isotopes escape from reactor buildings under pressure and once picked up by the wind, travel all over the world. Tellurium has no biological use and is both radioactive and poisonous.

Tritium: Half-life of 12.3 years. Beta emitter.

Tritium is an isotope of hydrogen, which means that when it joins up with another hydrogen atom and an oxygen atom to form a molecule of water, the "tritiated water" is indistinguishable from regular water and cannot be separated from it. This is causing quite a problem at the damaged Fukushima nuclear site, where they can get all the other radioisotopes out of the tons of contaminated water at the site, but not the tritium. Tritium is formed naturally from cosmic radiation. It is also formed in large amounts in atomic bomb blasts and in nuclear reactors. However, it is not considered a very significant radiation hazard.

Uranium-235: Half-life of 703.8 million years. Alpha emitter.

Uranium-235 is the uranium isotope capable of fission, and it is what "enriches" uranium-238 in order to make nuclear fuel and "weapons-grade" uranium. Various isotopes of uranium and its decay products are found in fallout.

Uranium-238: Half-life of 4.5 billion years. Alpha emitter.

Uranium-238 is the most common isotope of uranium found in nature, and is used to make plutonium. It is a common constituent of fallout, both from nuclear bombs and nuclear reactor accidents.

Zirconium-93: Half-life of 1.5 million years. Beta emitter.

Since zirconium is used in making the cladding that covers nuclear reactor fuel rods, there is plenty of it available in a nuclear reactor. Plenty is found in fallout from nuclear testing, as well as in reactor accidents.

Isotopes Made In Soil By A Nuclear Blast

In a ground burst, where an atomic bomb strikes the ground and blasts a large amount of soil and dirt high into the atmosphere, neutron bombardment of normal constituents of the soil can turn formerly innocent soil elements into radioactive versions of themselves. The following are some isotopes of interest from this process.

Aluminum-28: Half-life of 2.3 minutes. Beta emitter.

Aluminum-28 is formed from stable aluminum in the soil by neutron capture. It contributes to the high initial radiation in an area just following a nuclear explosion, but is mostly gone in an hour.

Carbon-14: Half-life of 5,700 years. Beta-only emitter.

Carbon-14 is formed naturally when cosmic radiation strikes stable nitrogen in the atmosphere and a neutron displaces a proton. In an atomic-bomb blast, lots of carbon-14 is formed from the surrounding atmosphere and soil. Carbon-14 gets into the human body usually as carbon-dioxide gas, where it is incorporated into tissues. It is a large contributor to the overall radionuclide count in fallout from atomic bombs around the world.

Manganese-56: Half-life of 2.6 hours. High-energy gamma emitter.

Stable manganese in the soil under neutron bombardment from an atomic explosion can "capture" a neutron and become an unstable isotope of manganese. It contributes very penetrating gamma rays to the scene just after a nuclear explosion.

Silicon-31: Half-life of 2.6 hours. Beta emitter with some gamma rays.

Silicon, a common stable constituent of soil and sand, when under neutron bombardment from an atomic bomb blast, can capture a neutron and become the unstable isotope, silicon-31. Again, this radioisotope of silicon would contribute chiefly to the radiation at the scene in the hours immediately following the bomb blast.

Sodium-24: Half-life of 15 hours. Beta and high-energy gamma emitter.

Sodium, common in both the soil and in sea water, undergoes neutron bombardment in a ground burst or a sea burst. Sodium-24 is formed by neutron capture in appreciable but highly variable amounts after a nuclear explosion. Given the fairly short half-life, most of it would decay within a couple of days after an explosion.

These are just a few of the almost 200 radioactive isotopes that are formed from nuclear bombs and in nuclear reactors. The ones that are of most interest to us are those that contribute most to the high amounts of radiation in fallout and those that enter into our—and our animals'—food supplies.

Remember that although most of these isotopes seem to be beta radiation emitters, most beta and alpha emitting isotopes also emit some to many gamma rays as they decay and rearrange their energies.

4: *Your Radiological Instruments*

The only way you can know there are radioactive isotopes included in whatever is falling from the sky is to have a radiological instrument that can measure it.

In our experience, you absolutely need only three pieces of radiological equipment: a survey meter, a dosimeter and a dosimeter charger. If you want to add to your capabilities, you can never have too many meters!

Radiation cannot be detected with any human sense organ. For this reason, if you feel that a nuclear attack on the United States is likely, or if you live anywhere near a nuclear power plant that could undergo a terrorist attack or an accident, then you might need a radiological meter on hand.

Many fine instruments are available on the market, and some are relatively inexpensive. Digital meters abound, and these are fine instruments, provided they measure what you are interested in.

How Meters Work

Most radiological meters today work by measuring how many ions are produced in gas when the gas is struck by radiation. When radiation strikes a gas molecule, it knocks off electrons and leaves a positively charged ion and one or more negatively charged electrons.

Scientists took advantage of this process and developed means of collecting and measuring the charged particles created by radiation. The stronger the radiation, the more charged particles are produced and the higher the reading your meter shows.

What Are You Interested In Measuring?

When you say you are interested in measuring radiation, this has multiple meanings to the manufacturers of radiation meters. This is because meters available to the public, that are affordable, usually measure either high levels of radiation or low levels of radiation, and you must choose a meter that fits the use you intend to make of it. There are meters that will measure both very, very low and very, very high levels of radiation, but they are usually unavailable and unaffordable to the general public.

Low-Level Meters

A meter that measures low levels of radiation is what you want if you intend to measure contamination in your foods, or in a fish you caught in the pond by the nuclear power plant. It is also useful when fallout is just beginning, or when fallout has almost totally decayed.

Most of the newer digital meters out there are low level meters. They do not tell you this. You have to know what you are looking at, otherwise, you may wind up with a meter that will be almost useless to you in a real fallout situation. If the meter you are considering measures in "microsieverts" or "milliroentgens," you are looking at a low-level meter.

The problem with these meters in a fallout situation is that their ranges are too low to measure the radiation levels found in fallout. If you check the meter's range, you will usually discover that it will measure no higher than something like 0.1 roentgen, or 0.5 roentgen.

Fallout can range from low levels as it is just arriving to double or triple digits when it is fresh and heavy. A low-level meter will "zero out" or go off-scale immediately in a radiation field like this. All you will know is that you are in a radiation field that is stronger than 0.1 roentgen, or whatever the top of your meter's scale happens to read.

The good thing about some of the newer low-level digital meters is their extreme sensitivity to very low levels of radiation. Some will measure as little as 0.001 milli-roentgen/hour, which is the sensitivity you want if you are trying to measure radiation levels in foodstuffs.

High-Level Meters

In a fallout situation, you want a high-level meter – something that measures more than 1 roentgen, and preferably all the way up to something like 500 roentgens. Fallout measurements can be quite high in radiation, especially fallout less than 24-hours old.

As *Radiation Safety in Shelters* puts it: *The levels of radiation from fallout from nuclear weapons can be much higher than those encountered in peacetime situations. The radiation instruments developed for use by operators of nuclear reactors, by radiation therapists in hospitals, or by crewmen of nuclear submarines and ships are not generally suitable for the needs of people caught in the radioactive fallout of a nuclear war. These commercial instruments for peacetime purposes usually do not have the higher ranges which may be needed for wartime use.*

This is why meters used by hazmat responders and by workers around nuclear reactors, although they may have higher ranges than most low-level meters available to the public, may still have ranges too low to adequately measure levels of radiation in fallout.

Digital high-level meters can be relatively expensive. For this reason, we usually recommend one of the old Civil Defense meters that measures from 0 to 500 roentgens. These are analog meters (they utilize a scale and a needle to indicate measurements) that are very hardy and were built to withstand a certain amount of rough treatment.

In particular, the CD V-715 is a great all-round meter that will measure higher levels of radiation from 0 to 500 roentgens. It makes use of four ranges, x0.1, x1, x10 and x100, which means you can measure in hundredths of roentgens at the lower levels.

The CD V-715 High-Range Meter

However, if your aim is to measure low levels of radiation found in foods or on people, you would want a meter that can measure very low levels, such as milliroentgens. The CD V-715 can often tell you there is radiation present at low levels, but it lacks the sensitivity to measure accurately the very low levels found in foods.

Radiation Measurement Systems

Scientists today live in a publish-or-perish world, and they are always in search of something new and different to write about and get published. Nowhere is this more evident than in the field of radiation measurement.

The roentgen, named for William Roentgen, the discoverer of X-rays, was among the first units of radiation measurement. It measures the amount of ionization produced in the air by X-rays or gamma rays.

When scientists discovered that radiation damaged human tissues, they developed terms to measure the amounts of radiation that would produce certain amounts of damage. The *rad* is a measurement of the radiation absorbed by a material or tissue. The *rem* is a measurement of the biological effect of the absorbed radiation. For all practical purposes, the roentgen, the rad and the rem are considered equivalent.

These units of radiation measurement are part of what is called the *Roentgen System* of radiation measurement.

Even though this is the system everyone is most familiar with, somebody out there always thinks he has a better way of doing things.

Someone came up with the *sievert*, a new unit of radiation measurement that supposedly takes into account the biological effects of radiation on tissue that is named for Rolf Sievert, a Swedish radiation physicist who studied the biological effects of radiation.

Nobody seems quite sure how this is better, but somehow everyone agrees that this is a better measurement than roentgens, so in the 1970s, the international scientific community switched from the roentgen system to the

Sievert, or International System, also known as the SI System.

Interestingly, the only difference that means anything to most of us is the fact that 1 sievert is equal to 100 roentgens, and herein lies the danger.

Many modern meters read in both sieverts and roentgens, which means you had better choose one system of measurement and stick with it. Otherwise, you run the risk of believing you're in a 5-roentgen field of radiation, only to learn that the person who measured it was using the Sievert System, and you are actually in a 500-roentgen field.

"Five" doesn't sound like much radiation, does it? If it's 5 roentgens, it isn't. But if it's 5 sieverts, that's a high radiation field, a 500-roentgen field, one that can kill you if you spend an hour or so in it.

Because of this hundred-fold difference, and the difference it can make to your health, we recommend sticking with the Roentgen System. Most of the good charts and literature that tell you what you need to know about levels of radiation are done in roentgens.

In a fallout situation, stick with what you know and what is familiar to you. It is not a good time to decide you are going to switch to the International System.

Learn To Use Your Meter

Once you have chosen a meter, spend a little time familiarizing yourself with it. Learn how to use it and practice using it to "survey" an area. The CD V-715 is called a *survey meter* for this reason – it is used to survey an area for radiation.

If you are able to afford a second meter, by all means get a low-level meter for use in surveying or "frisking" people for contamination, or for measuring the levels of radiation in foods or water.

If you like the Civil Defense line of meters, the CD V-700 is an excellent low-level meter that was intended for use in checking out people for contamination or for use when radiation levels had dropped to very low levels.

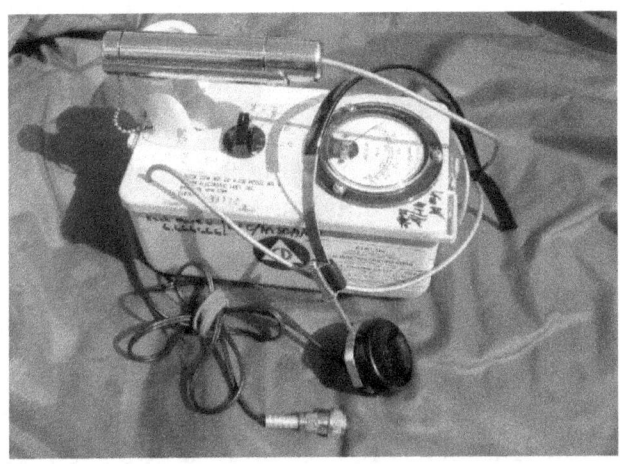

The CD V-700 Low-Range Meter

Be sure that the meter you choose has been calibrated. Many people buy a Civil Defense meter off eBay and assume it is in prime condition. It may or may not be, especially considering that some 40 years have gone by since these meters were last calibrated. If you own or obtain such a meter, be sure and have it calibrated by the experts at ki4u.com.

You may also want to obtain a "source," a low-level radioactive item you can use to check that your meters are working.

Once you have a meter and know how to use it, be sure to take background radiation readings of various areas of your home and land and write them down in a notebook. These readings will serve as your comparison levels if fallout arrives in your area.

Take background radiation readings in your garden and in your animal shelters and poultry yard, even inside your chicken coop. These readings will help you determine if your land or buildings are contaminated after an incident.

A radiation reading over twice your normal background level is considered contamination.

Dosimeters & A Charger

If you wish to keep track of how much radiation your body has absorbed, you need an instrument called a *dosimeter*. The job of a dosimeter is to sum up all the radiation you are exposed to so that it can be entered into your health records in case it should become important later.

The most inexpensive and easiest to use dosimeters are probably the CD V-742 pen dosimeters that you can buy in quantity from eBay. They read from 0 to 200 roentgens. One end of the dosimeter is used for charging the instrument. The other end contains a magnifying lens you use to view the scale. The Civil Defense dosimeters are about the size of a fountain pen and feature a clip like that on a fountain pen to clip the instrument to your pocket.

Depending upon the strength of the radiation field, you may be able to use your dosimeter CD V-742 without recharging it for the entire length of the incident.

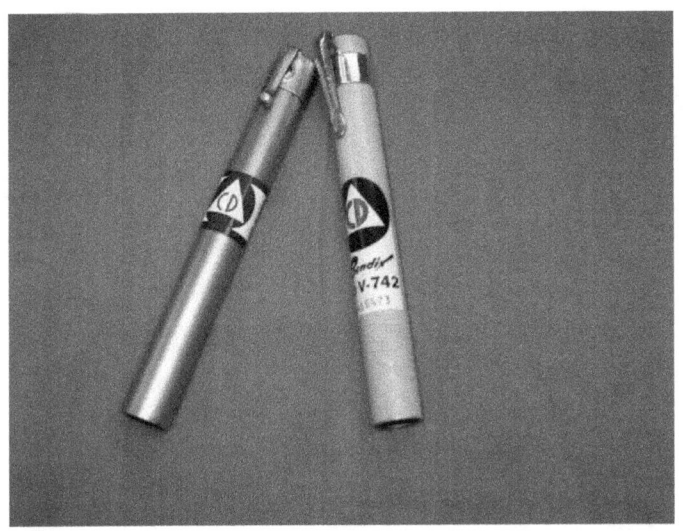

CD V-742 Dosimeters

In order to "charge" your dosimeter, or set the scale to zero, you will need a CD V-750 dosimeter charger. This is a square metal box powered by a single D-cell battery that charges a dosimeter so that you can set the scale to zero. One charger can charge many dosimeters, so you usually only need one charger, no matter how many dosimeters you have.

You will probably want each person on your property to have a dosimeter, and if you have extra, you can hang them in spots around your shelter to monitor the radiation dose received in that particular area.

Be aware that if you hang a dosimeter near a wall, the reading may be somewhat lower than the reading your personal dosimeter displays. This is because the wall acts as a partial shield so that some radiation does not reach the dosimeter.

If you want to be sure you have a working charger and dosimeters, especially if you are not gifted in working with electronics, you should order them from one of the companies that obtains its Civil Defense equipment from ki4u.com, such as http://store.advancedmart.com/. Then buy more dosimeters cheaply from eBay.

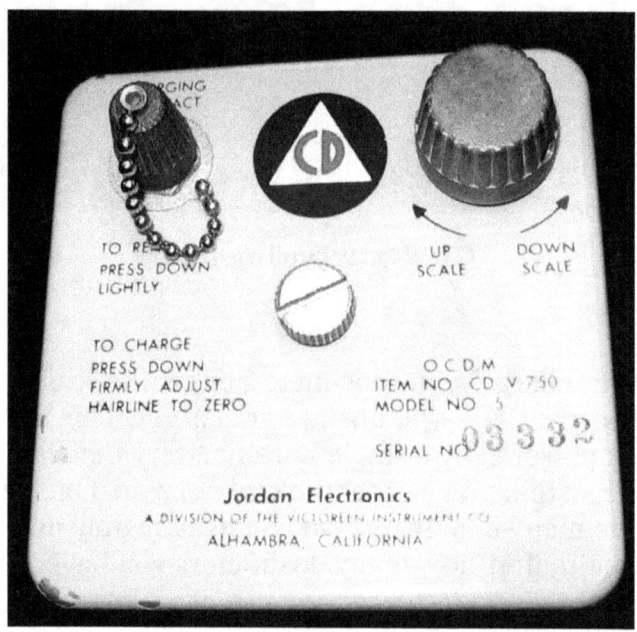

The CD V-750 Dosimeter Charger

For more information on how to choose radiological equipment, especially from the Civil Defense line, see Book I of our Radiation Series, *How To Choose A Civil Defense Radiological Instrument*.

This book also gives basic instructions on how to operate the Civil Defense equipment, and a description of the meters, dosimeters and chargers available.

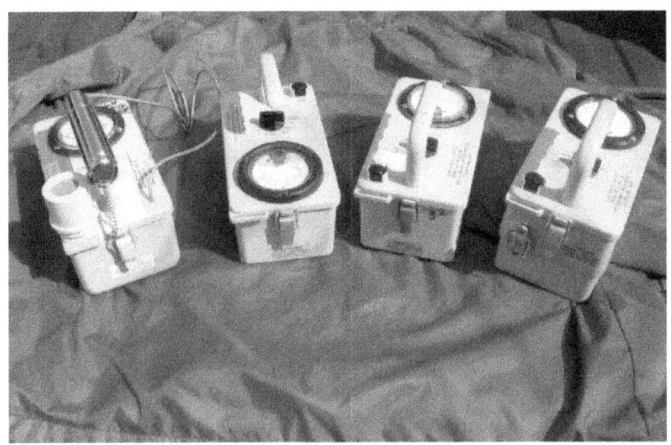

The Civil Defense Meters

CD V-700, CD V-715, CD V-720, CD V-717

5: *Your Home Fallout Shelter*

If you suspect that a nuclear incident may be in the future of your particular area, you want to give some thought to where you and your family will take shelter if fallout should come your way.

In protecting yourself and your family from radiation, you will make use of the three principles: *Time, Distance,* and *Shielding.*

We have already seen that radiation decays with *time*, and that is why you plan on taking shelter for a week to two weeks, in order to allow the fallout time to decay to low levels of radiation that pose no harm.

Radiation strength also drops with *distance.* As you increase the distance between you and a source of radiation, the strength of the radiation drops inversely by the square of that distance. This is called the Inverse Square Law, and we make use of it by seeking a place to shelter that is as far from fallout as we can manage.

Radiation also weakens as it travels through mass, so we will use as much mass as possible as *shielding* between us and the radiation. As we have seen, alpha radiation is stopped by a sheet of paper, and beta radiation is stopped by a sheet of tin foil. Gamma radiation, however, is pure energy and as such is very penetrating. The more mass, in the form of brick, concrete, soil, wood, books, furniture, etc., between us and the radiation, the better we are shielded from gamma rays.

Choosing Your Shelter Site

You may have an excellent area already in existence on your property. Some homes have a basement or a brick-lined room as part of the building. Some farms have a suitable inner room, brick-lined, in one of the outbuildings. If so, then plan on using this area for your family fallout shelter.

If you have no obviously suitable area available, then you can plan your shelter for an inner room in your home.

The ideal room for a fallout shelter is a center room of the house, preferably one without windows. Inside this room, you can plan to construct, out of furniture, file cabinets, and book shelves, a small structure that is just big enough to hold you and your family ... and your pets. The "roof" can be a door or a big slab of plywood that you will pile high with totes full of books, bags full of clothes, stacks of old magazines or newspapers, or even water containers full of stored water.

There are as many methods of constructing so-called "expedient" fallout shelters as there are locations. Some people may have suitable areas outside their homes where they can dig a deep trench, roof it with sheets of plywood, and pile it high with soil.

You can even build a "lean-to" shelter against your home and pile the side high with soil.

The important thing in constructing a fallout shelter is to put as much mass between you and the radiation as possible. The more furniture or storage totes full of books or family heirlooms you can stack around and on top of it, the better. If you have bags of soil you can pile high around it, that would be even better. Just be sure your "roof" is well-supported and capable of holding the mass you stack on top of it.

Staying Cool Or Warm

Unfortunately, we do not get to choose when the nuclear event will occur, and depending upon the time of year you need to use your home fallout shelter, heat and cold may be more of a concern than you planned for. A nuclear event may mean there is no electricity and no running water, which complicates your plans considerably.

In summer, excess heat is a major concern. Without air conditioning, and with windows closed to keep out as much fallout as possible, people can feel really bad and irritable as the summer heat rises. Hand fans and enough water to sponge off occasionally can make a big difference in attitudes.

Candles during a hot summer add even more heat to the atmosphere, so be sure you have stored enough batteries to keep your flashlights powered.

In winter, if heat if not available, you must utilize blankets for warmth. This is a time when candles will add some welcome heat to the air.

What To Eat And Drink

A nuclear attack upon America would no doubt result in an absence of electricity, possibly for quite some time. This would mean many of your appliances will not work, and there would be no running water. This is why you should have enough potable water stored to supply each member of your family with 2 quarts of water per day for two weeks.

In a case like this, the more water you can store, the better, because in the hot summer you will need extra water for sponging off in order to cool down.

If your stove is not working, you may have a propane grill or camping store that you can use for cooking, but in the hot summer, a stove will add unwelcome heat. In the winter, the heat from the stove might be welcome, but you do not want your family exposed to any carbon monoxide fumes. Use the stove only in a well-ventilated room.

This is why the Office of Civil Defense once stored crackers, hard candy and water for the occupants of the public shelters. The crackers and hard candies do not require cooking, and their digestion does not require prodigious amounts of water. The fallout shelter diet was planned to provide somewhere around 900 calories per day, enough to stay functional but definitely not a luxurious diet.

Plan your meals, either of stored canned goods or other items that require no cooking. For as long as you remain in your home fallout shelter, plan to eat foods you have stored that require no cooking. If you should have cooking capabilities, then your shelter life will be a little more luxurious, but this is something you should not count on.

Sanitation

During a shelter stay, especially if there is no running water, sanitation becomes twice as important as it normally is. Americans are used to running water and the freedom from many diseases it provides. For this reason, you will have to be twice as careful to avoid cross-contamination when there is no running water.

You must either have a portable emergency toilet on hand or use your own toilet. Without a fairly large source of water, such as a swimming pool, you will not be able to flush your toilet. But you can convert it to a usable emergency toilet by placing plastic bags in the bowl and discarding the bags after a certain number of uses.

Keep Lysol or some other disinfectant on hand for use in the toilet.

Cleanliness is a must during your shelter stay. Do not allow trash or food scraps to lie around on the floors or sit on the cabinets, and definitely do not allow flies in the vicinity. This is a time when you will have trouble dealing with disease in any family member, and that's assuming you can find a doctor.

Tracking Your Radiation Dose

Once fallout begins coming down, each member of your family should don his or her own dosimeter and keep it on at all times. Each day, you should record each person's radiation reading.

If you should have to go outside your shelter for a short time to perform an essential duty, keep a note of the radiation dose you receive during that time outside.

The best way is to maintain a chart for each family member. When the incident is over, you will know how much radiation each person received. This may become important in the future when decisions must be made about certain jobs and medical treatments.

Radiation is believed to be cumulative, and each person responds to radiation differently. Many people who received a heavy one-time dose of radiation at Hiroshima and Nagasaki never developed any problems. Other people who received a much lesser dose developed related health problems. *It depends on the individual.*

As you can see, many things must be considered when planning to spend a week or two in a home fallout shelter. For in-depth information on this subject, see Book II of our

Radiation Series, *Your Home Fallout Shelter: How To Ensure Your Family's Health & Survival In A Nuclear Incident.*

Now that you know the basics of what to do for yourself and your family, let us next consider how we can best protect our animals and our food sources, including our home gardens, from the hazards of radioactive fallout.

6: *How To Prepare For Fallout*

When you think of preparing for fallout, think of an approaching dust storm. Whatever you would do to protect yourself and your belongings from breathing in or being covered with dust is what you will do to prepare for the arrival of fallout, whether in the form of nuclear-bomb debris or a radioactive plume from a nuclear power plant.

You may have no warning of approaching fallout, only to discover that it has already arrived and there is little you can do to prepare.

But if you should have any time at all to prepare, whether because of an announcement or by some other means, or you suspect that something is about to happen, then begin at once to enact your preparations.

Think about this now and make a list of items you will need to cover or put into shelter. Store the list in a place where you can find it instantly. Once you get such an announcement, you are likely to forget half the things you need to do because of the stress of the moment.

Yourself & Your Family

If you have any warning that a nuclear event is occurring or about to occur, your first obligation is to your family. You should begin at once to prepare your home fallout shelter to protect your family. See *Your Home Fallout Shelter: How To Ensure Your Family's Health & Survival In A Nuclear Incident* for advice on how to do this.

Your Pets

If you know that fallout is coming, or if there is even a possibility that fallout will come to your area, get your pets inside at once.

With cats and dogs, the primary thing you do not want is for fallout to get on or into their fur. The best way to prevent this is to get them inside or into some other form of shelter that will keep the fallout off them.

Do not trust your dog to stay in his dog house during the arrival of fallout if you keep him outside. The only way to really protect your dog is to keep him inside your own house, or in some shelter like a barn that will keep all fallout from landing on his fur. You do not want your pets to suffer beta burns on their backs or anywhere else on their bodies where fallout can collect.

Cats are known to groom themselves by licking their fur regularly, and although they may clean fallout off their fur, you certainly do not want them swallowing it and irradiating themselves from inside.

The Alamogordo Cattle

Interestingly, the major incident that alerted officials to the problems of fallout and range animals occurred accidentally. On July 14, 1945, the United States tested the world's first nuclear bomb in the desert near Alamogordo, New Mexico.

At around the same time, a group of Hereford cattle on a nearby ranch managed to escape their pasture and wandered onto the government land, just in time to receive fallout deposits from the bomb blast on their backs. Then they returned to their own pasture, where the ranchers didn't notice anything unusual for a week or two.

The damage consisted of a loss of hair and burn scarring down the backs of the cattle. Patches of hair grew back discolored white. Scientists soon realized the animals had received beta burns because the fallout remained on their backs for some time. The Army wound up buying some 88 of these fallout-marked cattle from ranchers and studying them over their lifetimes in hopes of learning more about the long-term effects of fallout.

One of the major conclusions appears to be that livestock tended to recover remarkably well, and their overall lifetime productivity was not affected.

Your Livestock

If you can avoid any beta-burn damage to your animals, by all means do so. Whether you have one milk cow or a herd; one goat or twenty, or if you keep some pigs or other farm animals, if it is at all possible, get them inside a shelter. Even a pole barn is better than being outside, because it will keep fallout from depositing on their backs. It even offers a little protection from radiation in general.

Range animals that are grazing on the range may not be so easy to shelter, but if you have the time and it is at all possible, round up your animals and get them to shelter. If it is not possible to shelter them all, studies show that if you can get them beneath a grove of trees, the trees provide some shelter from fallout.

Failing a grove of trees, you can corral them in an outdoor pen. Although it won't shield them from fallout as it is coming down, the animals crowding together will help shield each other from radiation. Plus, if you have them in a pen, it will be easier to quickly remove any fallout that has deposited on their backs.

Russian studies advised farmers to put a blanket over the backs of their animals if fallout was expected. If you have only a few cows and no available shelter for them, this might be a viable option for you. In this case, any cloth or plastic you can get on their backs to keep fallout off their bodies while it is actively coming down would be helpful in protecting their skin from beta burns.

Experts advise that if you own dairy cattle or goats, they should be given the most protected areas. This is because of the danger of radioactive iodine finding its way into the dairy animals, and thus into their milk.

If it is hard to shelter all your animals in time, shelter the most valuable animals first.

As for poultry, even though you may hate to shut your backyard chicken flock into their coop, this is the one thing you can do that will protect them the most. Chickens tend to remove fallout from their feathers fairly rapidly, and they tend to be far more resistant to radiation in general than other livestock, but the more radiation you can help them avoid, the better they will do in the long run.

Animal Food Supplies

If at all possible, cover all exposed animal food supplies. The less contamination your animals ingest, the better for their health.

Much animal food is kept in covered bins, silos, or hay lofts. You can add to the protection by putting a tarp over any food storage container. Be sure and close any windows or vents that might allow fallout (think dust!) into your animal food storage areas.

If you have a haystack or bales of hay that are kept outside, cover them as thoroughly as possible with tarps or plastic sheeting. Even old sheets or quilts or blankets will help keep fallout off your animals' food supplies.

Generally, dog food or cat food is kept inside your home in covered containers. If this is the case, then it is adequately protected already. However, if you keep any dog or cat food where fallout can get on the container, either cover the container or move it to a location where fallout can't get on it.

Animal Water Supplies

If at all possible, uncovered animal water supplies should be covered with tarps or plastic sheeting. If your animals usually drink from a pond, it will not be possible to cover the pond, so your best bet would be to plan alternate water sources for your animals until any fallout that lands in the pond has a chance to decay.

Running water such as a stream or a river, even though fallout may land in it, tends to stay fairly clear of contamination because of the moving water.

Your Garden

Your vegetable garden should be covered if at all possible with plastic sheeting or a tarp. Leave the tarp over the plants until you are sure fallout has stopped being deposited. This is the best way to keep your garden vegetables free and clear of radioactive contamination, and to protect your plants from being burned or from taking radioactive isotopes into the plant itself.

The same should be done for any fruit trees you are cultivating. Plants are especially susceptible to radiation when they are actively forming fruit. Radiation can reduce their yields considerably.

Experiments done during the nuclear testing of the 1950s show that a simple sheet of polyethylene plastic spread over plants gave an amazing amount of protection to the plants sheltered beneath it.

If you have a garden or maintain outdoor watering troughs for your animals, or any other water source that is important to people and animals, be sure and keep a store of tarps and/or plastic sheeting, of the type used to cover floors and furniture when painting, on hand at all times.

You never know when these supplies will come in handy.

7: *What If Fallout Arrives Unexpectedly?*

The way things tend to happen in this world, you may have little to no warning that a nuclear incident is about to happen. Your first warning may be arrival of what looks like grit, dust and ash that appears on smooth and shiny items, like your car.

If you are not expecting a dust storm in your area, check your local news stations. If all the news stations you can usually receive are no longer on the air, and if your power has mysteriously gone off, suspect the worst.

Take shelter inside your house immediately, and call in your pets. Brush them down well on the porch or other anteroom of your home then take them to the inner rooms. Wear a mask or a respirator and gloves and goggles while you brush them down.

Brush yourself down as well if you have any of the dust or grit on your body or in your hair. Keep as much of the radioactive dust and grit in this outer room as possible where it can easily be swept up and thrown out. Treat fallout as you would dirt, but handle it only while masked and gloved.

If at all possible, wash your pets if they have been outside in the fallout, even though you have brushed them well. Washing will not deactivate fallout, but it will remove it to another location. You do not want it in your pets' hair for any length of time as it may give them beta burns.

As you saw from the list of isotopes present in fallout, most of them seem to be beta emitters, and this is why you do not want fallout collecting on your pets or your livestock.

Fallout is heaviest and most radioactive in the day or two following the bomb blast, because, as we have seen, many of the radioactive atoms decay within a few hours. That means you must shelter yourself, your family, and your pets for several days to two weeks, until the fallout has decayed to safe levels. And there is only one way to know what the levels of radiation are at any given time.

Get Out Your Radiological Meter!

The only way you can know for sure if what is coming down on your homestead is radioactive is to check out your surroundings with your own meter. If you suspect that the gritty, sandy stuff that's coating everything could be fallout from a bomb blast on the big city a hundred or so miles away from you, or that something bad has occurred at the nuclear power plant near you, get your meter out immediately and find out.

Your meter should be stored, without batteries, if you haven't been expecting an incident. Quickly install fresh batteries and turn the instrument on. The Civil Defense CD-V meters usually require a warm-up period of a minute or two.

While your meter warms up, grab one of your zeroed dosimeters and clip it to your pocket or neckline. If there is fallout, you want to begin tracking your radiation dose now.

Generally, the meter you begin your measurements with is your high-range meter. In the Civil Defense lineup of instruments, this would include the CD V-715, the CD V-717, and the CD V-720. Zero the meter as instructed then perform an operational check. If the meter checks out okay, then check radiation levels at the nearest window.

The best way to begin taking readings is, after the meter is warmed up, set the range selector switch to X0.1, and if the needle goes off-scale to the right, go to the next range, until the meter no longer goes off-scale to the right.

The reading you obtain will determine your next actions. At certain low radiation dose rates, you can go outside your shelter to perform certain tasks, and how long you stay outside depends upon the dose rate. At higher dose rates, you should not go outside your shelter at all. It is essential to give your body time to heal and repair the radiation damage it has already sustained.

Precautions For Outside Operations In Radiation

Less than 0.5 R/hr: Essential precautions should be observed, such as respirator mask, goggles, wide-brimmed hat, full-length raincoat down to your ankles, boots with their tops up under the bottom of the raincoat, whiskbroom and gloves. You can only stay out in this for a few hours per day. *You should sleep in the shelter.* You must give your exposed body time for healing and repairing!

0.5 R/hr to 2 R/hr: Same precautions as above. Activities should be restricted to essential duties, such as put out a fire; secure food and water; acquire medical supplies; save a life. *Do not stay out more than two hours per any one day*!

2 – 10 R/hr: Hold time outside the shelter for no more than a few minutes a day for absolutely essential purposes!

Greater than 100 R/hr: *No outside activities permitted*!

The above precautions were designed for people staying in fallout shelters, so they would have an idea of how much work they can do outside the shelter if the necessity should arise during the time they are staying inside their home fallout shelters.

The main thing to remember in a situation where you have to go outside the shelter during a time of fallout is that your body must be given the chance to repair the damage from radiation. This is why you can work outside several hours day when radiation is less than 0.5 R/hr, but you still must sleep inside your shelter.

If The Reading Is Very Low ...

If your meter indicates anything below 1 or 2 roentgens, you will be able to perform some actions to mitigate the effects of the fallout on your animals. If your CD V-715 shows a reading of less than 1 roentgen, you may wish to use a low-range meter, if you have one, to obtain a more exact reading.

If you have a low-range CD V-700 meter, and you believe the reading is somewhere below 0.5 roentgens per hour, get out your CD V-700 or other low range meter and find out what the exact reading is. For low readings, while fallout is just beginning to arrive, a low-range meter will give you a more exact reading.

Get your family busy preparing to enter into your home fallout shelter. Have them fill any available container with water, in case water supplies are compromised, and cover any foodstuffs that are sitting out, such as bowls of fresh fruit.

For a brief period, you may be able to take some preventive actions with your livestock. But before you go outside, put on a wide-brimmed hat covered with plastic wrap or foil (for ease of cleaning), gloves, boots and an overcoat or a raincoat. Also put on a face mask, preferably a respirator, and wrap-around goggles, or use the RediMask that combines an eye shield with a respirator. When you go out, take your meter with you inside a plastic bag. You do not want your meter contaminated with fallout.

Once outside, you can quickly cover any animal watering troughs, animal food or your vegetable garden with the plastic sheeting you have hopefully stored up, or with a tarp.

Depending on how many animals you have and where they are located, you may have time to get them into shelter. If you have animals in a pasture far from your home, you may have to leave them out, but any animal that is near your barn, get it inside the shelter as quickly as possible. Brush the animals off with a long broom as a safety precaution.

While you are working, periodically check your meter. Fallout can come down rapidly and the radiation levels can rise quickly. Once your meter registers over 0.5 roentgens per hour, you should complete your preparations and hurry back to your fallout shelter. You should not be out in fallout with radiation greater than 0.5 roentgens per hour for longer than two hours on any one day.

If the radiation is less than 0.5 roentgens per hour, do not remain outside in fallout any longer than necessary. Quickly do what you can to gather your livestock and poultry into their shelters, cover what food and water supplies you can, then get back to your family fallout shelter.

Take another reading to see if the radiation has risen, and if so, how much. If the radiation has gone above 0.5 roentgens per hour, you know fallout is continuing to come down and the radiation levels will likely rise even more.

Once you are back inside, shed all your outer clothing in an anteroom and use a whiskbroom to brush down any areas on your body that may have picked up some fallout. You want to bring as little fallout as possible into your inner home fallout shelter.

If The Reading Is Already High ...

If your meter registers 0.5 roentgens or above, your best decision is to remain inside your home fallout shelter. Help your family in the preparations, and when completed, all of you, including your pets, should move into the shelter area.

If absolutely necessary and the radiation readings are below 2 roentgens per hour, you may be able to get a few of your animals that are close to their shelter inside, but be sure to wear a raincoat, boots, gloves and a broad-brimmed hat, with goggles and a respirator while doing so. Do whatever you can do outside to cover your garden and animal food sources while remaining close to your shelter. Don't remain outside for longer than 2 hours.

If fallout is actively coming down, radiation will likely continue to increase. You can take readings every 10 or 15 minutes to monitor the rise. Do not remain outside in radiation above 2 R/hour for more than a few minutes per day.

Once fallout reaches levels of 2 to 10 roentgens per hour, you should remain inside your shelter. The only time you should leave your shelter is for something absolutely essential, such as to save a life or bring in supplies.

Once fallout levels are greater than 100 roentgens per hour, you should not go outside your shelter for any reason for any amount of time.

If You Should Need To Lie On The Ground ...

In the course of making your last minute preparations as fallout is actively coming down, it may become necessary for you to lie on the ground to rescue a person or an animal, or perform some necessary task on your land. Be aware that fallout's ultimate destination is the ground, and this is where you will come in contact with more of it.

As you noticed in the list of isotopes in Chapter 3, many of them are beta emitters. Their half-lives may be short, but while fallout is coming down, many of those with short half-lives are still viable. In other words, the freshest fallout is the most dangerous, and if you have to crawl on the ground to rescue someone, you will be exposed to even more.

If you should have to perform a rescue or some other operation that requires you to be on the ground during a time of fallout, be sure and stand up every so often and brush yourself off. If you're trying to drag someone to safety, stop and brush him off, also. If beta-emitting particles are allowed to remain on the skin, that is when the damage can occur. Your job is to make sure those particles don't remain on your skin for long. Do not under any circumstances allow fallout to remain on your skin for longer than a few hours. Less is better.

Most people who have received beta burns from fallout were burned because they simply did not know they should get the stuff off their skin as soon as possible.

This is also why you should keep your head covered when you are outside in fallout. You do not want "dust" getting into your hair and on your scalp at a time when water may be in short supply and you won't be able to wash.

If you wear hair dressings or pomades, or special hairstyles involving braids, covering the head is especially important. If you should get fallout dust stuck to your scalp or hair or within your braids, your best bet would be to cut off the hair if you are unable to wash it out because of a lack of water. This may be a consideration with long hair, also. You do not want beta burns on your scalp at any cost!

8: *Protecting Your Pets*

As we have already seen, the best way to protect your pets, especially dogs and cats, is to get them inside with you, into your home fallout shelter. Utilizing the principles of *time*, *distance* and *shielding* is the Number One way to protect any living creature from radiation dangers.

Dogs and cats, like people, need special places so they can relieve themselves. Dogs can often use a room in your home that is not being used for your fallout shelter. Spread old newspapers or a tarp over the floor. Your dog will usually be happy to be inside with you during the two weeks or so you will need to be inside your shelter.

Cats, if trained to use a litter box, should be fine kept indoors for two weeks, since many cats are kept inside all the time. You should have no problem not allowing your cat outside for around two weeks.

The big problem will be letting your pets outside briefly, or if the pet decides it wants outside at all costs and makes a dash for the door. You should avoid this if at all possible, as your pet will be contaminated upon return to the house. In this case, you would have to brush each pet down well in an outer room before allowing them back inside the shelter. Then survey each pet's fur with your meter to be sure you've removed all contamination.

Giving each pet a bath before letting them back inside would be even better, but it is all too likely that water will be at a premium during this time.

Should You Give Your Pets KI?

Once you know that fallout is coming down, it is time for your family to begin taking the potassium iodide that you have stored. The first few days, as we have seen, is the most critical for radioactive iodine, as the half-life of I-131 is 8 days.

Potassium iodide, given in high doses, is known to protect the thyroid glands of human from radioactive iodine. The thyroid gland selectively snags iodine from the bloodstream, and it cannot tell any difference between radioactive iodine and stable iodine, so it will readily snatch up radioactive iodine. Radioactive iodine irradiates the thyroid and destroys thyroid tissue, perhaps even causing thyroid cancer to develop years later.

Many pet owners, knowing this, rush to give their pets potassium iodide even before fallout arrives, often gauging the dose by weight of the pet, as you would gauge it to the weight of a child or a baby. The problem is that dogs and cats are not human beings, and they may not process iodine in the same way as humans.

Some websites dedicated to pets advised that your pet should be dosed with potassium iodide according to weight, and the weights and doses used are human doses.

However, there are veterinarians who warn against this, as the side effects can be rough on the pet if it gets too much. And the problem is that nobody really knows how much is too much, as all the studies have been done in humans. It is entirely possible that dogs and cats process iodine differently than humans, in which case the doses given to humans would be entirely different from the doses given pets.

At high enough levels in pets, too much potassium iodide can cause death. Less rigorous side effects include allergic reactions, vomiting, diarrhea, loss of appetite, depression of thyroid function and heart damage.

Because it is bitter, potassium iodide can cause nausea and salivation. At high doses, as during a time of fallout, it can cause tearing, irritation of mucous membranes, cough, dry coat and hair loss. It has been associated with cardiomyopathy in cats.

One veterinary oncologist, Dr. Michael Lucroy, during the time the Fukushima disaster was at the forefront of everyone's consciousness, said, "If you're giving commercial pet food, plenty of iodine is in those diets, anyway."

Another Vet's Opinion

Dr. Al Wehner, located in Beaumont, Texas, however, wonders what dog food Dr. Lucroy is speaking of, as he is not aware of any commercial dog food that is particularly high in iodine.

Curious, we decided to check on this. We discovered that most commercial dog foods, such as Alpo, claim to contain iodine, but they do not specify how much. So it appears to us that Dr. Wehner is right; commercial dog food is not what we'd call high in iodine. As a veterinary oncologist, perhaps Dr. Lucroy has access to special "cancer diet" pet foods that the rest of us are unaware even exist.

Dr. Wehner also points out that the *Saunders Handbook of Veterinary Drugs*, a vast compendium of animal treatments, documents potassium iodide as a treatment for pets' thyroid glands in case of a nuclear event in addition to other uses.

Dr. Wehner says flatly, "If something like a nuclear attack with fallout were to happen, and there is nothing else that you could do, I would dose my pets as given in the *Handbook*. In a case like that, what have you got to lose?"

If you suspect that a nuclear event may occur, and you choose not to give your pets potassium iodide, you can begin supplementing your pets' diets with kelp powder, which is a safe iodine supplement. By the time an event occurs, your pets would hopefully have gradually saturated their thyroids with stable iodine.

An important thing to remember is that you, your family and your pets should not begin taking KI until an attack occurs. Do not give KI as a preventative, just in case an incident occurs. You do not want to risk the side effects possible with heavy KI dosage unless there is a genuine reason.

Suggested KI Oral Doses For Small Animals

Small Dogs (up to 20 lbs): 15 mg/day

Medium Dogs (20-50 lbs): 65 mg/day

Large Dogs (Above 50 lbs): 100 mg/day

Cats: 10 mg/day

9: *Handling Contamination*

If fallout comes upon you suddenly and heavily, your best bet is always to take shelter at once and take any pets that are outside into the shelter with you.

You should remain in your home fallout shelter until radiation has decayed sufficiently to allow you to do a few outside chores.

Consider fallout as you would dust. Wherever dust can go, fallout can also go. Keep this in mind when you venture forth to salvage whatever animal foods and garden produce that you can.

This is also why you wear a mask or respirator when you first go outside after fallout has decayed to safe levels. As you walk, the finer fallout particles may aerosolize again as you kick them up, just as dust will. You do not want to breathe any radioactive contaminants into your lungs.

If there has been no rain since the fallout came down, then you must be doubly careful of kicking up dust.

Rain, however, creates its own problems. It washes fallout off many surfaces, but it tends to pool in other areas, such that radiation levels in those areas may spike to higher levels. If you have any concerns about an area, check it with your meter.

Contaminated Animal Feed

If fallout has come upon you suddenly and you have not had time to cover your hay or animal food bins, all is not lost. Often, you can simply remove the top layer (very carefully!)

and the food beneath will be relatively free of fallout contamination.

Many experts even advise farmers that they can keep livestock penned for several days without food, provided the animals are given plenty of water. Depending upon the circumstances, this may be the best idea if your area is heavily contaminated by fallout.

The first 72 hours after fallout arrives are the most crucial. If you are caught with no animal food on hand that is stored so that it won't become contaminated, then give your animals plenty of uncontaminated water but no food for that length of time.

However, if you are in a situation where this is not possible, it is better that your animals receive contaminated food and water than no food or water at all.

Foods Stored in Silos or Bins

Animal foods that were stored in silos or covered bins should be safe for animals, provided there are no cracks or crevices for any "dust" or fallout to drift in and contaminate the foods.

If your outside silos or covered bins have cracks or crevices that could admit fallout, now is the time to correct the matter. Either caulk them or cover them with tar paper or tin now, before an incident occurs.

Feed Stored Outside

If you were not able to cover animal foods stored outside, such as hay bales, they can still be used. Carefully remove the outer layer of hay, usually 3 to 6 inches thick, and discard it. Wear goggles and a mask while doing so. The hay beneath will be mostly free of contaminants and can be safely used.

Food stored in uncovered bins is usually safe if you carefully remove the top few inches, wearing mask and gloves while you do so.

Animal feed in burlap bags that are stacked outside can be used if the top layer of bags is removed. Other types of bags can be brushed off or wiped off safely.

Do not discard the top layer of bags or the top few inches of grains that are stored in bins. If you have a meter, check the levels of radioactivity. Then store them. Chances are, when you check them a few weeks later, the radioactive contaminants will have decayed to much lower levels so that the feed can be used if necessary.

How long this feed should be stored before it decays depends upon what fission products came down and how much of them came down. The only way you can know for certain is by checking the feed with your meter.

Contaminated Water

Water sitting out in the open, such as a pond or a trough, should be considered contaminated after a nuclear incident. Do not use it if at all possible. Running water, however, such as a stream, is usually fairly safe because the contamination is swept downstream and spread out as it falls.

This is why you should prepare a means to cover any water storage containers that are kept in the open, such as troughs, for your animals. Once these containers are contaminated, about the only way to decontaminate them is to empty them out and scrub them out with uncontaminated water. This may not be possible if water is not running after an incident.

Be aware that water cannot be decontaminated by boiling it. This will kill bacteria and other germs, but it will not do a thing to radioactive isotopes. They must be filtered out, and one of the best ways to do this is to pour the water through a filter that contains a lot of crumbled-up clay. The radioactive isotopes are attracted to the clay and adhere to it. Activated charcoal does not do a good job of filtering out radioactive particulates, and neither does sand.

See:

http://www.chemicalbiological.net/water%20purification%20schematic.html

You can also distill the water, which is a fuel-intensive, time-consuming process, but this will also remove the radioactive isotopes along with most other contaminants.

When it comes to water, your best bet is to have sufficient water stored in covered containers for your animals, and for yourself and your family.

Contaminated Animals

If you have range animals that were out on pasture during the incident, they have probably received a certain amount of fallout on their backs. Animals on the range that move around a lot tend to get rid of a certain amount of it, and rain immediately afterward would help remove even more. But animals can receive beta burns to their backs along the area where the fallout rests, as did the Alamogordo cattle.

These burns may not affect the cattle's production or life spans, but it definitely affects their skin, causing scarring and hairlessness. If you are able to quickly remove any fallout from your range animals' backs, you may be able to prevent this.

If you have running water, you can wash any fallout deposits off your animals. Wash them down in one area then have the animals walk through a broad, shallow pan of water in another area to remove any fallout from their hooves.

Failing this, use a broom or other long-handled brush to brush them down as much as possible. Be sure you are wearing a respirator, goggles and other protection from fallout while you do so.

Washing or brushing simply removes fallout from your animals' backs and onto the surrounding ground. It does not stop the radioactivity. It merely moves it from one place to another. Be sure the area where you wash or brush your animals down is not an area you will be using for other activities.

This is why protecting your animals beneath a roof, or even in a forested area, to keep fallout from being deposited on their backs is your best choice, but if this is not possible, then remove any deposits off your animals as quickly as possible.

If you have enough available, or you own a colloidal silver generator, spray colloidal silver on your animals' backs if you believe the fallout deposits may have been there long enough to cause beta burns.

Can You Eat Your Own Produce?

If you were able to spread tarps or plastic over most of your garden, your produce should be safe to eat. Wash the vegetables well and prepare them as usual. Experiments have proven that gardens covered like this are remarkably free of contamination. Check to be sure with your low-level meter.

If your vegetable garden was covered with fallout, however, wash the garden down thoroughly with a water hose if possible. Then whatever produce you harvest, wash it thoroughly and test it with your meter before using.

Fallout is like dirt. Once you physically wash it off, it is usually gone. Some isotopes can be absorbed by certain plant leaves, but generally this is a small amount. Discard the wash water carefully to avoid contaminating anything else.

If the vegetable or fruit can be peeled, then wash it and peel it and discard the peelings. Do this very carefully so as to avoid contaminating the insides of the fruit with the knife that is removing the peeling. Treat the peelings as radioactive waste and dispose of them where animals can't get at them.

Items such as peas and beans should be washed then shelled, and the shells discarded. Corn can be washed and husked, and the husks discarded. The vegetable inside should be free of contamination. Check with your low-level meter to be sure.

Greens, such as mustard or turnip greens, are an "iffy" item. You can wash them thoroughly, but greens can absorb some radioactive isotopes through the leaves. In particular, strontium-89, strontium-90 and ruthenium-106 can be absorbed into leaves and tend to remain in the leaves. Cesium-137 and iodine-131, if absorbed by the leaves, will then move throughout the entire plant.

Once you have washed the greens thoroughly, check them with your low-level radiological meter. If the reading is around twice the normal background radiation level of your garden, do not eat the greens.

Remember: *Cooking does not destroy radioactive isotopes*! If they have been taken into the plant leaves themselves, you will not be able to wash it away or otherwise "disable" it.

Root vegetables are safe to eat immediately after a nuclear incident. Wash them well and prepare as usual. They have not had time to absorb any radioactive isotopes, especially if there has been no rain since the incident.

Fresh fallout is usually surface contamination only. Your job is to remove that surface contamination without contaminating anything else in the process.

As time goes by, some isotopes may be absorbed into the leaves or the plant. That is when you want to have a low-level meter capable of detecting the low levels of contamination in food plants you hope to eat.

Some contaminated food items may be safe to eat if you are able to store them for a long enough period of time so that the radioactive contaminants have time to decay. You can freeze some foods. Others can be dried and placed in storage. But wait several weeks before testing them with your meter to see if they have become safe to eat.

Milk

Milk is a special topic of its own because of radioactive iodine. When fallout from either a nuclear event or a nuclear plant accident lands on pastureland where milk cattle or milk goats graze, the animals ingest radioactive iodine and other isotopes with the forage. Radioactive iodine is passed directly from the animal into the animal's milk.

This is only a problem for about a month or two after the incident because radioactive iodine decays fairly rapidly. Iodine-131 has a half-life of only 8 days, and iodine-133 has an even shorter half-life of 22 hours, less than one day. But the milk produced by animals on the range immediately after a nuclear incident should not be used as is. Research has shown that 5 to 10% of radioactive iodine consumed by cattle is passed through into their milk.

The reason for all the concern over radioactive iodine in milk is that young children drink the most milk, and because of their youth, they are more at risk of thyroid cancer and other problems that can be caused by ingesting radioactive iodine. If some of your milk animals are outside during a fallout event, bear this in mind, especially if you have children who usually drink the milk your animals produce.

Do not allow children to drink any milk that may be contaminated. Experts even suggest that if a nuclear attack is possible and you have children, to store powdered or canned milk for this reason.

Some producers discard the milk produced during the critical period immediately after an atomic blast. However, this is not necessary if you have the capability of freezing the milk and storing it for long enough for the radioactive iodine to decay.

Other producers use the milk to make cheese or other milk products, which can then be stored until the radioactive iodine has decayed. Cheese can also be stored until some of the longer-lived isotopes have had a chance to decay.

After a nuclear event, always test foods before eating or drinking, including those you have put aside in storage so that any radioactive isotopes in them can decay.

Guidance For Farmers

Most of the literature concerning livestock and other farm products reiterates that if you are a farmer, you will be told what to do with your livestock and animal products, and you will be given suggestions for handling contamination in your pastures and fields.

If this should prove possible in this day and age of people who are terrified of radiation and all its mysterious power, then you may actually receive some pertinent advice on certain aspects of your animal or farm products. Some states have produced booklets on the possibility of a radiological incident via their agricultural departments that advise farmers and other producers that they will receive advice on when products may be safely sold and when badly injured animals may be put down, etc.

However, given that an all-out nuclear attack may well result in a shutting down of the electrical grid, *don't count on it*. These authorities will likely be fighting to preserve their own lives, and even if they have advice to give, they may well not have the technology available to give any advice. If the electrical grid is down, so will be the internet and most phone service.

If your state agricultural department has such a booklet available, see if you can obtain a copy. The more information you have available in the case of a radiation incident, the better prepared you will be. Often, these booklets give you hints as to what kind of advice you can look for from the authorities of your state—provided they have the means available to deliver it.

But always be ready to act on your own if these authorities are unavailable. That is why you are reading this book!

10: Using Your Animal Products

If you keep farm animals for milk production, meat or eggs, will you be able to use them yourself in the event of a nuclear attack or a nuclear plant accident in your vicinity?

Often, you can safely do so, provided you take precautions to avoid contamination. This is especially true if your area is heavily contaminated.

Some products can be safely stored until the radioactive contaminants in them have had a chance to decay.

But you will not know this unless (1) your local authorities tell you when it is safe to sell or use certain animal products or (2) you have a meter capable of giving you that information.

You should also know the background radiation of your kitchen before you begin the cooking process. Foods tend to register very low levels of radiation unless they are heavily contaminated. Background radiation, before any fallout has arrived, is from cosmic radiation and the naturally occurring radioactive components in the ground. What you will be looking for is whether or not the foods register about *twice your usual background radiation*.

Assuming a major attack, and that you have no way of selling your products, you (and perhaps your neighbors) will likely need to eat or use them yourselves. If so, here is how to safely do that.

Sick or Dying Animals

It is possible that some of your animals may be affected by the radiation in fallout so much that they become very sick, or they may even die. If it is possible for you to treat the animals, you should do so. Many animals can recover from radiation sickness and live productive, normal lives.

However, if you see that an animal cannot be saved, you should euthanize it. Most of the literature insists that you wait until you receive instructions from the agricultural authorities before you dispose of the animal's body. But in a real nuclear crisis, you are not likely to hear anything from these authorities, even if the lines of communication are up.

If one or more of your animals dies, do not wait around to hear from the authorities. Bury the animal. Never eat an animal that is sick or has died. Do not leave a dead animal unburied, because other animals will eat it and spread the illness or contamination.

Eggs

Eggs are usually safe to eat, even if your chickens have consumed contaminated feed. Generally, however, chicken feed is kept stored in bins and will not be contaminated.

The major contaminant found in eggs will usually be strontium-89 and strontium-90, which behave like calcium. In the case of eggs, most of the contamination will be found in the egg shells, which are composed largely of calcium.

One way to find out is to use your meter to measure the contamination in the whole eggs, shells and all. Then crack the eggs into a bowl, being careful not to allow any bits of shell to fall into the bowl with the eggs.

Measure the egg shells then measure the eggs. Chances are the eggs in the bowl are relatively free of radioactive contamination, while the shells register a certain amount of radioactivity. This is likely radioactive strontium contained in the eggshells. Discard or bury the shells where animals can't get at them.

If your bowl of eggs measures some radioactivity, search the bowl carefully for bits of eggshell. Remove any bits you find and rescan the bowl with your meter.

Poultry

If any of your poultry behaves in a sickly manner, allow them to recover. Do not eat any bird that is sick.

If your family wants a cooked chicken for dinner, sacrifice the chicken in the usual way and skin it. Do not bother to pluck it, simply skin it and discard the skin with the feathers. Remove the organs and discard them. Do this while wearing a mask and gloves in case the chicken has collected some radioactive dust in its feathers.

Do not eat any of the internal organs, not even the liver or the gizzard. Do not eat the feet, as they may have been in contact with radioactive contaminants enough to possibly absorb some.

Remove the bones, as any radioactive strontium will usually be found there Or you can cook the chicken with the bones in and remove the bones easily after cooking. Cook the chicken in a pressure cooker if you have one, or boil or bake it as usual. Rinse meat thoroughly before measuring it with your low-level radiological meter.

Gather the bones into a pile. Using your low-level radiological meter, measure the radioactivity of the bone. You can also measure the radioactivity of the inner organs, skin and feathers. Most of the radioactive strontium will be concentrated in the bones. Most other isotopes will likely concentrate in the organs and skin. The meat, if it contains any radioactive isotopes at all, will likely contain very little.

Red Meat

Do not eat any animal that is not completely well. If any animal acts as if it is sick, allow it to recover.

The animal should be slaughtered and skinned carefully in order to avoid contaminating the meat with any radioactive debris on the skin or in the hair. The organs should not be eaten, as most radioactive contaminants tend to collect in certain organs. Very few radioactive isotopes are found in the muscle meats. Do not eat the bone marrow or chew on the bones.

If you have concerns about radioactivity, use your meter to check the levels in the muscle meats you intend to eat. Also check the organs (which you will not eat) and the skin. Chances are you will see that most radioactivity remains in the skin and in the organs.

The bones (and the bone marrow) should not be eaten. Radioactive strontium tends to concentrate in bones, so simply cook the meat and bone as usual then remove the meat from the bones. Rinse the meat well in a colander. Place it on a separate platter and measure for radiation. All you should detect is background radiation.

11: *Restoring Your Garden*

How much use you can make of your garden after fallout descends upon it depends on what stage your produce had reached before the fallout arrived. For instance, if fallout comes down on a garden that is just starting to flower, the plants may be damaged so badly that they produce very little. Generally, plants are most vulnerable to radiation when they are in the flowering stage.

If your fruit trees or produce was ready to harvest, you may possibly lose that fruit or produce because you are unable to get outside and harvest it. Do not go outside when radiation levels are high just to harvest your crops. Wait until radiation levels have decayed to safe levels as given in Chapter 7 before you attempt to salvage any of your produce.

Much produce, however, can be harvested as usual once it is safe to do so, provided you take precautions *against breathing in fallout, or getting it on your skin.*

Decontaminating Your Garden

Once you have harvested what you can harvest from your garden, it is time to remove what you can of the radioactive contaminants from your garden.

First, wearing a mask and gloves, pull up all the plants in your garden and dispose of them in an area as far away from your garden as you can find. Make sure your animals can't get to it, because many will consider garden plants fine eating.

Liming, or adding lime to soils that are naturally low in calcium will do much to reduce the uptake of radioactive strontium by plants. Since strontium competes with calcium for absorption, if you make sure there is plenty of calcium available, plants will preferentially absorb calcium rather than strontium.

Soil that is naturally low in potassium can have wood ash or a commercial source of potassium like Greensand added. This may help prevent the uptake of cesium-137.

One of the most recommended methods of decontamination of soil is to literally remove the top layer of soil and deposit it in a place where it is away from people and animals.

If you are determined enough and your garden is small enough, you can accomplish this on your own with a shovel and a wheelbarrow.

Failing this, you can turn the soil over, by digging or plowing deep enough to put the top layer of soil below the root levels of the plants you intend to grow. That way, the radioactive isotopes are transferred below the root level of your plants, and fresh soil from below is brought to the surface.

Grains

Some homesteads may grow fields of grain for consumption by their livestock. Strontium absorption usually is not a problem with grains, nor is cesium, because grains naturally are low in both calcium and potassium.

If you grow grains, contamination usually is not a problem in the first place because they don't tend to absorb radioactive isotopes. Plus contamination on grains is on the outer seed coat, so milling and polishing usually removes most of it.

Pastureland

In a garden, scraping off the top layer or plowing deep might be feasible, although contaminated pastureland is usually too much acreage to either dig deep or to scrape off the top layer. If fallout is light, the pasture might be usable as it is. Natural decay of the isotopes and the action of the weather often reduces contamination levels considerably within a few weeks.

If fallout is heavy, agricultural experts recommend harvesting the hay or forage and disposing of it elsewhere, then plowing the land deep and reseeding it. If you have sufficient lime available, you can also lime the land to cut down on plant absorption of radioactive strontium.

Remember, strontium-90 is a bone seeker, and the less of it your animals absorb, the less you and others who consume your animal products will absorb.

What To Plant In Case Of Future Fallout

What if fallout is not a problem yet, but you suspect that in the near future it might be? Are there some items you can plant that will be less vulnerable to the isotopes likely to be in fallout?

Much study has been done on plants and what isotopes they may pick up or absorb. In general, alfalfa, clover, soybeans and leafy vegetables are more likely to absorb radioactive strontium than fruits, grains and potatoes.

Bear in mind that when it comes to plants, even with the wide range of sensitivity to radiation they display, the environment also influences their sensitivity greatly. Environmental factors most important in influencing the radiosensitivity of plants are light, temperature and competition. So if all these environmental factors are perfect, radiation effects on plants will likely be much less.

Other studies indicate that wheat, oats and barley are very sensitive to beta radiation; corn, potatoes and sugar beets are moderately beta sensitive, and rice is very resistant to beta radiation. Beta radiation can actually burn plants at certain stages. Since beta radiation is largely given off in early fallout, the effects on plants will be much worse from heavy fallout blown in from a nearby explosion.

Many of the effects of radiation on plants have to do with what state of development the plant is in at the time of exposure. For instance, legumes appear to be most sensitive to radiation during their flower-bud stages.

Of the root crops exposed to radiation, radishes developed a bad taste and poor texture when they were irradiated at higher exposures. Sugar beets had reduced sugar content, and garlic had considerably reduced yields.

Of the pasture and forage crops, irradiation of sweet clover reduced growth almost 80%.

For The Home Vegetable Garden

If you think fallout might be coming your way and you want to plant things you will be most able to harvest after fallout has come down, consider planting the following.

Brussels sprouts

Cabbage

Canteloup

Carrot

Cucumber

Garlic

Green Beans

Legumes: Beans and Peas

Okra

Onions

Potatoes

Potatoes, sweet

Squash, acorn

Squash, zucchini

Squash, winter

Squash, yellow

Watermelon

Yams

Many people enjoy greens such as mustard, collard and tender greens. Plant them if you have the extra space, but remember that these must be well-washed if a radioactive fallout incident occurs. Also, they can absorb some radioactive isotopes through the leaves.

The same goes for lettuce, especially the leafy varieties. Headed varieties are usually better for a fallout situation because you can peel off the outer leaves and remove most of the contamination.

If you save seed for next year's crop, be aware that radiation may affect the vigor and viability of a plant's seed. The seed may look normal and you may not see the effects until after planting them the following year.

12: *Radiation Effects on Insects & Worms*

Since earthworms and insects have such a big effect on agriculture, studies have been done on how they react to radiation.

Insects

Big agriculture these days is basically a monoculture, and this makes things really good for insect populations that feed on certain plants. The huge farms of today tend to raise only one enormous crop of a single grain or vegetable, whereas the farms before World War II cultivated a variety of vegetables and livestock.

Therefore, the big farms of today tend to be plagued with certain species of insect pests that are specific to that food being cultivated. Crop losses to insect pests amount to billions of dollars annually, and the cost of chemicals to control the pests amounts to millions.

For this reason, we hoped we could look forward to a little bit of good from a fallout incident. Maybe the insect populations would be wiped out, and the crops would be left alone for a few years.

Alas, this is not quite the case. In fact, it seems that radiation can be considered a pretty fair insecticide, in that the pests may be eliminated from certain areas. However, they appear to get going again immediately, just as they do after a good application of insect killer.

In fact, radiation differentiates among insect species much like an insecticide, even though insects as a group are fairly tolerant of radiation. Some species are affected by a certain dose more than others.

The problem is that most of the insect pest species are very widely distributed, and when you wipe an area clean of them, others from outside areas begin moving into the clean area right away.

Researchers who investigated the question of insects and radiation, probably sorry to have to give us the bad news, point out that even if an area is wiped clean of a certain species, other representatives of that species would be moving in from outside the area before the radiation had even cleared.

In other words, if your area is hit by radiation, you can expect pest insect control to be a problem in trying to get your garden going again, even if you found every pest insect in your garden feet-up on the ground.

Sometimes, there is no justice.

Earthworms

We were not surprised to discover that earthworms had been studied in regard to their response to radiation. They are known to be extremely important to agriculture because of their ability to turn over and aerate the soil, and because of their ability to convert organic matter into plant foods.

The news here is all good—earthworms are among the most resistant of the soil's inhabitants to radiation. Scientists expect few earthworms to die from radiation.

Three reasons explain this:

1. Earthworms naturally have a high resistance to radiation.
2. The soil is an excellent shield against radiation. The deeper the worm is in the soil, the more it is shielded.

3. Radioactive particles have time to decay before they are incorporated deeper into the soil.

Laboratory experiments showed that earthworms in certain locales, such as a forest, can rapidly turn over forest leaf litter, and that the quantity of organic matter turned over is dependent solely on the amount available rather than the earthworms' ability to deal with it.

This means that in certain areas with plenty of organic matter and lots of earthworm activity, soil will be turned over fairly rapidly, and radioactive particles that had been deposited on top would soon be worked deeper into the soil. This is analogous to plowing the top layer of soil deep below plant roots.

What this suggests to us is that if you have cultivated a garden with plenty of earthworms, you can expect a lot of help from them in turning over your garden soil. This could save you a tremendous amount of heavy labor.

Therefore, our suggestions for prepping a garden for the possibility of radioactive fallout are:

1. Make sure your soil has a high level of calcium.
2. Make sure your soil has a high level of potassium.
3. Make sure your soil has a high level of earthworms.

And remember to keep plenty of tarps or plastic sheeting on hand. If the worst should happen, covering your garden is one of your best protections.

Radiation Safety Limits

0 to 50 Roentgens	Considered Safe
50 to 200 Roentgens	Level I Radiation Sickness
200 to 450 Roentgens	Level II Radiation Sickness
450 to 600 Roentgens	Level III Radiation Sickness

Level I Radiation Sickness: Less than 5% deaths. From 5% to 30% of exposed people may develop acute symptoms of nausea and vomiting within 4 hours of exposure. A temporary reduction in blood platelets and white blood cells may occur.

Level II Radiation Sickness: Less than 50% deaths. From 60% to 75% of exposed people may develop acute symptoms nausea and vomiting within 4 hours of exposure. Severe blood changes, hemorrhage, and hair loss.

Level III Radiation Sickness: Greater than 50% deaths. One-hundred percent of exposed persons will develop acute symptoms of nausea and vomiting within 4 hours of exposure. Severe blood damage, hemorrhage, and hair loss, with up to 80% deaths in less than 2 months.

Generally speaking, the higher the dose of radiation received, the faster the onset of symptoms. At Chernobyl, the doses received were so high, firefighters noted a metallic taste in their mouths and a severe headache before they developed nausea and vomiting that forced them to stop their work.

In a couple of "criticality accidents" in experimental laboratories (cases where an unexpected event caused a brief, high release of radiation), the radiation-blasted person immediately felt as if he was burning and tingling all over and developed almost instant nausea and profuse vomiting.

As has been said by many experts in the field, if the radiation dose is high enough for you to actually feel, it's also high enough to kill you.

Here is a small chart you can copy onto a 3 x 5 card and keep with your meter. In an actual incident, stress tends to degrade your memory, and radiation is something you do not want to take chances with.

1 Roentgen = 1 Rem = 1 Rad = 0.01 Gray = 0.01 Sievert

0 – 30,000 CPM = Safe

Greater than 30,000 CPM = Questionable

```
1R = 1Rem = 1Rad = 0.01Gy (Gray) = 0.01 Sv (Sievert)
0 - 30,000 CPM = Safe
Greater Than (>) 30,000 CPM = Questionable
```

Radiation Limits For Outside Activities

Less than 0.5 R/hr: Essential precautions should be observed, such as respirator mask, goggles, wide-brimmed hat, full-length raincoat down to your ankles, boots with their tops up under the bottom of the raincoat, whiskbroom and gloves. You can only stay out in this for a few hours per day. *You should sleep in the shelter.* You must give your exposed body time for healing and repairing!

0.5 R/hr to 2 R/hr: Same precautions as above. Activities should be restricted to essential duties, such as put out a fire; secure food and water; acquire medical supplies; save a life. *Do not stay out more than two hours per any one day!*

2 – 10 R/hr: Hold time outside the shelter for no more than a few minutes a day for absolutely essential purposes!

Greater than 100 R/hr: *No outside activities permitted*!

The above precautions were designed for people staying in fallout shelters, so they would have an idea of how much work they can do outside the shelter if the necessity should arise during the time they are staying inside their home fallout shelters.

The main thing to remember in a situation where you have to go outside the shelter during a time of fallout is that *your body must be given the chance to repair the damage from radiation*. This is why you can work outside several hours day when radiation is less than 0.5 R/hr, but *you still must sleep inside your shelter*.

Resources

Actions For Survival: Saving Lives in the Immediate Hours After Release of Radioactive or Other Toxic Agents, 2011, Allen Brodsky

A Field Guide to Radiation, 2012, Wayne Biddle

Chernobyl 01:23:40: The Incredible True Story of The World's Worst Nuclear Disaster, 2016: Andrew Leatherbarrow

Defense Against Radioactive Fallout on the Farm, 1958, U.S. Department of Agriculture

Emergency Information for Farmers, Food Processors and Distributors, New Hampshire Department of Safety

Instruction and Maintenance Manual: Radiological Survey Meter (CD V-715), 1962: The Victoreen Instrument Company

Instruction and Maintenance Manual: Radiological Survey Meter (CD V-717), 1964: The Victoreen Instrument Company

Introduction to Radiological Monitoring: A Programmed Home Study Course, HS-3, 1972: Staff College, Defense Civil Preparedness Agency

Operating and Maintenance Instructions: Radiological Dosimeter Jordan Model 750-5, Jordan Electronics, a Division of the Victoreen Instrument Company

PLoS One, *Whole-Body Counter Evaluation of Internal Radioactive Cesium in Dogs and Cats Exposed to the Fukushima Nuclear Disaster*, Wada D, Ito N, Watanabe M, Kakizaki T, Natsuhori M, Kawamata J, Urayama Y, January 2017

Possible Effects of Nuclear Radiation Accidents on Agriculture, Bell MC, Riechert BP, 1987, University of Tennessee Agriculture Experiment Station

Protection of Food and Agriculture Against Nuclear Attack: A Guide for Agricultural Leaders, Agriculture Handbook No. 234, U.S. Department of Agriculture, 1962

Radiation Safety In Shelters, Federal Emergency Management Agency, 1983

Radiological Defense; Textbook, 1963, United States Department of Defense, Office of Civil Defense

Radiological Emergency Manual for Livestock, Poultry, and Animal Products, 1987, Berger CD, Frazier JR, Greene RT, Thomas BR, Auxier JA, IT Corporation/Radiological Sciences Laboratory, Oak Ridge, TN

Saunders Handbook of Veterinary Drugs: Large and Small Animal, 3rd Edition, Mark G. Papich, 2010

Science, *Barium-140, Radioactivity in Foods*, Anderson EC, Schuch RL, Risher WR, Van Dilla MA, February, 1958

Survival of Food Crops and Livestock in the Event of Nuclear War, *Proceedings of a Symposium Held at Brookhaven National Laboratory*, Bensen DW, Sparrow AH, Editors, U.S. Atomic Energy Commission, December, 1971

The Effects of Nuclear Weapons, 3rd Edition, 1977, Glasstone S, Dolan, PJ, U.S. Department of Defense and Energy Research & Development Administration

About the Author

Dr. Charles S. Brocato is a scientist and author who has written extensively in the fields of surviving chemical and biological warfare and nuclear warfare. He is also a longtime martial arts instructor, weapons/firearms instructor, and a map & compass instructor, specializing in *how to stay alive.* He is also a nutritionist, a counselor in the field of nutrition and an award-winning French chef.

He holds degrees in biology and mathematics with advanced studies in the areas of biochemistry, microbiology, and advanced mathematics, and a D.D. (Doctor of Divinity). He has also completed numerous studies in areas of Emergency Management with FEMA and MetEd.

Two of his previous books are available on Amazon in print form: *The Two-Fold Chastisement: Visions of the Coming Earth Changes*, and *Chemical/Biological WarFare: How You Can Survive*.

This book is the fourth in a series of Radiation titles he has planned in the area of surviving a nuclear attack or incident.

Write him at: **csbauthor@chemicalbiological.net**

Other Books By This Author

Dr. "B"s Radiation Series

Book I: *How To Choose A Civil Defense Radiological Instrument: Geiger Counters & Dosimeters*

Book II: *Your Home Fallout Shelter: How To Ensure Your Family's Health & Survival In A Nuclear Incident*

Book III: *Nutrients Against Radiation Damage & Injury In A Nuclear Event: Something You Can Do Before It's Too Late*

Book IV: *How To Protect Your Pets, Livestock and Home Garden In A Nuclear Event*

Other Topics

Rosemary: The Healing Herb of St. Martin de Porres

The Two-Fold Chastisement: Visions of the Coming Earth Changes

Chemical/Biological WarFare: How You Can Survive

Dr. "B"s Motto:
"You Can Never Have Too many Meters!"

www.ingramcontent.com/pod-product-compliance
Lightning Source LLC
Chambersburg PA
CBHW070152230526
45471CB00002B/629